PHAC

JUN - - 2021

SPOOKED

ALSO BY BARRY MEIER

PAIN KILLER: AN EMPIRE OF DECEIT AND THE
ORIGIN OF AMERICA'S OPIOID EPIDEMIC

MISSING MAN: THE AMERICAN SPY
WHO VANISHED IN IRAN

SPOOKED

THE TRUMP DOSSIER, BLACK CUBE, AND THE RISE OF PRIVATE SPIES

BARRY MEIER

HARPER

An Imprint of HarperCollinsPublishers

HarperCollins books may be purchased for educational, business, or sales promotional use. For information, please email the Special Markets Department at SPsales@harpercollins.com.

Originally published in Great Britain in 2021 by Sceptre, an imprint of Hodder & Stoughton.

FIRST U.S. EDITION

Designed by Nancy Singer

Library of Congress Cataloging-in-Publication Data has been applied for.

ISBN 978-0-06-295068-0

21 22 23 24 25 LSC 10 9 8 7 6 5 4 3 2 1

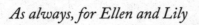

As always, for Ellen and Lily

CONTENTS

SPOOKED

STALKING MR. STEELE

FARNHAM, ENGLAND, 2019

There was a police car behind us so it seemed like the right time to ask Ian what we should do if they pulled us over. "It's good to get your story straight," he said.

"So what's your story going to be?" I asked him.

"I'm a driver," he replied.

That was true. He was sitting behind the wheel. But I had expected, given that he was a professional private detective, that his story would have covered me. "What should mine be?" I said.

"Well, you could be honest," he suggested.

Thankfully, we weren't stopped, but if we had been this is what I would have said: We were in Farnham stalking a resident, Christopher Steele, who once worked as a spy for MI6, Britain's equivalent of the CIA. We wanted to stake out his house but were having trouble finding it because the name of his street that we had found on a property database didn't match any names that came up in Google Maps. As a result, we had been driving around narrow country lanes in circles for an hour.

I wanted to meet Steele because of his role in the infamous

"dossier" about Donald Trump and Russia. During the 2016 presidential campaign, Fusion GPS, an investigative firm run by a former reporter for *The Wall Street Journal*, hired Steele on behalf of the Democratic Party to dig up information linking Trump to the Russians. Over a five-month period, Steele, who had spent four years in Moscow as an MI6 operative, wrote a series of memos that he sent to the ex-journalist, Glenn Simpson. Rumors about the memos had circulated for months within media and political circles before and just after the 2016 election. But the reports only publicly emerged in January 2017, two months after Trump's surprising victory, when BuzzFeed, an online news organization, posted them. Then all hell broke loose.

The memos, collectively known as the Trump dossier or the Steele dossier, went on to be cited in tens of thousands of articles, television programs, tweets, podcasts, and blog postings. Some of Steele's reports claimed that members of Trump's campaign had colluded with the Kremlin and one salacious memo suggested that the Russians had "kompromat," or comprising material, on the Republican candidate. It was a videotape, according to Steele's sources, showing prostitutes that Trump had hired peeing on a bed at the Ritz-Carlton hotel in Moscow once used by President Barack Obama.

Trump's foes embraced the Steele dossier as proof that the Russians had helped him steal the 2016 election from Hillary Clinton. The new president and his allies went on the attack, calling it "fake" news. Simpson and Steele became celebrities.

In the fall of 2019, Simpson and his partner at Fusion GPS, Peter Fritsch, who was also a former *Wall Street Journal* reporter, published a book titled *Crime in Progress*, which they described as telling the "inside story" of the dossier. Christopher Steele, who co-owned an investigative firm in London called Orbis Business

Intelligence, was enjoying the limelight, too. A Hollywood production company owned by the actor George Clooney had bought
the rights to his story, and in 2019, he attended a celebrity-filled
event at a trendy London restaurant. The party was held to honor
a new editor of *Vanity Fair* magazine, and guests included Colin
Firth, the well-known actor, and Monica Lewinsky, the former
White House intern with whom Bill Clinton had an affair and
who now wrote for *Vanity Fair*.

The magazine had earlier published a profile of Steele that depicted him as a kind of James Bond for the Trump era. It described
him as "an ex-spy who knew where all the bodies were buried in
Russia and who, as the wags liked to joke, had even buried some
of them." Steele was reveling in the moment. As the party was
breaking up, another guest asked him for his business card. Steele
thought he was being asked for his autograph so he picked up his
place card from the table and, pulling out a pen, signed it with a
flourish.

I was interested in Glenn Simpson and Christopher Steele for a
different reason. The dossier and the massive political and cultural
fallout it spawned epitomized the oversized impact that private
spies were suddenly having on politics, business, and our personal
lives. Around the same time that the Steele dossier was disclosed,
it was revealed that Hollywood producer Harvey Weinstein had
hired four different corporate investigations firms to dig up dirt
on women accusing him of sexual assault. One of those companies, Black Cube, which was based in Israel, dispatched a female
operative who posed as an activist for women's rights in order to
befriend one of Weinstein's accusers and get information that the
producer's lawyers could use against her.

Everywhere one looked, operatives-for-hire seemed to be

running amok. Reporters covering the trial of a lawsuit against a chemical company over the health dangers of its weed killer were approached by a woman who claimed to be a journalist but turned out to be working for a crisis-management firm representing the company. About the same time, a lawyer who long had been locked in a dispute with three businessmen from Eastern Europe discovered that private spies had planted a tiny, motion-sensitive video camera in a tree outside his home that recorded the license plate of every car entering the property. Elsewhere, private operatives working on behalf of Credit Suisse, a big bank, chased one of its former executives through the streets of Zurich, Switzerland. The incident began when the ex-employee discovered the spies were shadowing him and used his cellphone to take a picture of one of them. The private eye demanded that the man turn over his phone and, when he refused, the spy tried to rip it away. The chase ended with the arrest of an investigator.

Private investigators once were content to lurk in the shadows. Now, politicians were hiring them to dig up dirt on opponents, companies were employing them to torpedo investigations into their activities by authorities or journalists, and dictators were using them as freelance intelligence agents. A new generation of cheap, off-the-shelf technology was also making it easier for hired operatives to monitor cellphones, hack emails, and manipulate social media. Private spying was no longer a small business. It had become a hidden, billion-dollar industry. In the process, private spies had become more emboldened than ever before—just as their power to influence events had become more pervasive.

There is little question that private investigators take on legitimate assignments. They track down missing people, locate witnesses to testify in court cases, and conduct background investigations for companies into prospective executives or potential business partners. Still, everyone in the industry knows its secret—that the big money is made not by exposing the truth but by papering it over or concealing it.

Spies-for-hire are part of a wider web of enablers—lawyers, public relations executives, "crisis management" consultants—who serve the powerful and wealthy. But what makes private operatives unique is that they are the unseen part of that web, taking on the kinds of jobs that other people don't know how to do or don't want to get caught doing.

Private operatives invariably say they never engage in hacking. But instead of hacking themselves, some operatives farm out hacking jobs to subcontractors in India, Eastern Europe, or elsewhere. Private investigators will also say they do not misrepresent themselves in order to dupe the unsuspecting into revealing information. And some investigators don't. Still, much of the private spying world would become extinct if its inhabitants weren't willing to engage in activities that are illegal, unethical, or just plain unsavory.

While a reporter at *The New York Times*, I had gotten a glimpse into the world of hired operatives when writing a book about a former FBI agent turned private investigator named Robert Levinson, who disappeared in 2007 in Iran while working as a contractor for the CIA. The events surrounding the Steele dossier and Harvey Weinstein's use of corporate investigators were unfolding when I retired from the *Times* and it seemed like the right time to flip the script and investigate the investigations industry. I knew that scrutinizing private spies wouldn't change their behavior. Still, I wanted to understand how a predatory industry was operating unchecked.

Some private operatives spoke to me. Others wouldn't. When I approached Glenn Simpson of Fusion GPS about cooperating with this book, he wouldn't. Neither would his partner, Peter Fritsch. That was disappointing but at least they made their position clear. Over a period of six months, I also sent numerous inquiries to Christopher Steele, both directly and through his investigative firm in London, expressing my interest in interviewing him. He didn't acknowledge any of them.

That's where Ian Withers, my "driver" in Farnham, came in. A British journalist described Ian in a book about the infamous phone-hacking scandal at *The News of the World*, a tabloid owned by Rupert Murdoch, as "the best known and longest-surviving private investigator" in England. Ian's reputation could be summed up another way: in his prime, he had been a bad-ass who would take on any job and do whatever was needed to get it done.

We had first met in 2018 in New York at a conference of private investigators. Ian was seventy-nine at the time and had a round, florid face and a sizeable potbelly. A few months earlier, Scotland Yard had taken him into custody for questioning in a decades-old cold case murder. The victim, Gérard Hoarau, was assassinated in London in 1985. At the time, he had been seeking to overthrow the government of Seychelles, a tiny island nation off the coast of Africa, and Ian was a private CIA-style operative hired by that country's leaders to spy on him.

In 1986, the television newsmagazine *60 Minutes* aired an episode about Seychelles titled "Spies Island," which depicted the country as the epicenter of a geopolitical power struggle between the United States and the Soviet Union as well as a haven for money launderers. Ian was interviewed for the segment at a local restaurant he owned. Back then, he looked youthful and fit, though his tinted eyeglasses lent him a somewhat sinister appearance.

"Who killed Mr. Hoarau?" he was asked. "Who wanted him dead?"

"I cannot really tell you," he replied. "It could have been any government that had an interest in maintaining stability here."

The interest of British police in Ian as a source of information in the unsolved killing soon subsided. By then, we had gotten to know each other and when I asked him if he wanted to help me track down Christopher Steele, he said he was game. After spending a few days in London, I connected with Ian at Gatwick Airport, where we rented a car and drove to Farnham, a wealthy and bucolic suburb that lies about an hour west of London. The town has a quaint downtown, costly homes, and lush fields where sheep graze. I had tried to prepare myself for what lay ahead.

Jane Mayer, in a profile of Christopher Steele she wrote for *The New Yorker* magazine, described him as an ordinary-looking businessman who didn't stand apart from his fellow passengers on a commuter train save for a secret inside his briefcase. There, Mayer wrote, Steele kept his cellphones tucked inside special pouches called "Faraday" bags. The bags are made from a metallic material that blocks the signal a cellphone emits and so prevent bad guys from using them as a way to track a phone's owner. This sounded like something I needed, so I bought a Faraday bag before leaving New York. I imagined that Christopher Steele may have gotten his as a going-away present from MI6. I got mine from Amazon.

As we drove around Farnham, Ian explained to me that he often started stakeouts by determining whether his target was home. He would call the subject's house and pretend, if someone picked up the phone, to be calling from a taxi service. "Sir, I'm on my way, would you like me to pick you up outside?" he would say. When the person said they hadn't called for a cab, he would apologize for dialing the wrong number.

The surveillance strategy he used depended on how long a stakeout needed to last and how much a client was willing to pay. A top-of-the-line option at his agency was a surveillance van outfitted with night vision binoculars, a video camera, and a metal can in which an investigator on a stakeout could pee. The van was fitted with clips on either side that were used to hold inter-changeable signs advertising the services of an imaginary trades-man. One set of signs made it appear that the van belonged to a plumber, another set, an electrician, and still another set, a tele-vision repair service. The company names were different but the telephone number on all the signs was the same and it connected to Ian's agency. That way, if a suspicious homeowner or a neighbor called, a clerk who knew the disguise the van was using that day would answer appropriately. "Hello, this is Joe Bloggs plumbing, how can we help you?" Ian said, mimicking what a caller would hear.

Another ruse involved a set of props—a small tent, a set of plastic highway cones, and collapsible barricades—that made it appear that private eyes were workers doing road repairs. There were simpler methods. A private eye might park near a target's house, take out a bucket of water and soap, and pretend to wash his car. Another technique was to prop open the car's hood and pretend to be fixing it. This approach could be problematic, Ian said, because passersby liked to offer help. The most difficult type of surveillance, he explained, involved a home set at the end of a short dead-end street or another closed-in location that made it virtually impossible for a private detective not to get spotted. "What you don't like is a tight surveillance," he remarked.

Not long afterward, we finally found a small sign with the name of the street where Steele lived. It was immediately clear that this was going to be the tightest of tight surveillances. The lane was tiny and one house looked exactly like pictures of Steele's home I had seen.

When BuzzFeed posted the dossier, reporters for British tabloids had descended on Farnham. By then, Steele had fled and gone into hiding with his wife and children. "He asked me to look after his cat as he would be gone for a few days," one neighbor told a reporter. In newspaper articles, the large, two-story house was described as costing $2 million and equipped with security cameras mounted along its roofline. (To make sure readers didn't miss them, thoughtful editors drew circles around the cameras.)

Ian stopped the car and I got out. Some journalists like to surprise strangers by showing up at their doorsteps. I had never been one of them and there was a sense of relief when I saw that there weren't any cars parked behind the gate. As we drove away, Ian said that one way to stake out a house in a setting like Steele's was to have an operative lay hidden within bushes, concealed inside a camouflaged sleeping bag. He seemed to know the technique worked because he had used it.

Ian and I checked into a local inn and, later that evening, we went back to Steele's house. The driveway again was empty. Schools in England had just closed for summer vacation and it seemed likely that Steele and his family were away. We planned to drive the next morning to London to visit the offices of Orbis Business Intelligence and I decided to write a note to drop into Steele's mailbox on our way out of town.

While in London, I had gone to the restaurant, the Cora Pearl, that was the scene of the *Vanity Fair* party and taken a promotional postcard. It showed a portrait of the restaurant's namesake, a famous nineteenth-century courtesan who had luxurious curly brown hair and was wearing a low-cut gown.

"Dear Mr. Steele," I wrote on the card. "So sorry to miss you at the VF party, hoping to connect re my book on the P.I. industry."

The following morning, we drove back to the house. This time, four cars including a black Range Rover were parked behind the gate. I walked over to the mailbox and dropped the postcard into it. Then I pressed on a buzzer, which looked like it was equipped with a video camera. It rang five or six times. Nothing happened and it seemed like whoever was at home had decided to ignore me.

I was walking back to the car when Ian shouted, "The gate's opening." It was already rolling shut when I turned around but I managed to run through it. My anxiety grew as I approached the house. There was a large opaque window on the ground floor that looked like it might be part of the kitchen and I could make out the outline of a person behind it. When I got to the front door, it opened and Christopher Steele was standing there. I recognized him from a photograph that showed him dressed in a dark blue business suit, a crisp white shirt, and a tie. Now he was wearing plaid boxer shorts and a blue T-shirt. In the photograph, his graying hair was freshly coiffed and blow-dried. Currently, he had a serious case of bed-head and I presumed he had just woken up.

I explained the reason for my visit and asked if we could chat. "I can't talk today," he replied. "It's my birthday."

That threw me. I like to think I had the presence of mind to wish him "Happy Birthday" but can't be sure. I know I did ask, "Well, when is a good day to chat?"

"Why don't you send me an email," he said.

With that, our brief encounter was over. Back in the car,

I checked a database. It really was Steele's birthday and he had turned fifty-five. Later that day, I sent him a text message. "Happy Birthday. Trust it will be a wonderful one," my text read. "Once again, I am eager to chat with you about my book."

This time, he responded. "I'm sorry but I am completely log jammed with client meetings today and traveling from tomorrow," he wrote. "If you want to send some questions marked for my attention, I'll take a look at them next week. Chris."

A few written answers weren't what I had in mind. "Understood and thanks for your prompt reply," I replied. "I would appreciate if we could find a few hours." We never did. A year after our brief encounter, I sent Steele a list of questions. One of his colleagues replied, "We do not intend to respond."

The private investigations business is composed of a scattershot mix of people, drawn to the work by money, the opportunity for travel and adventure, and the heady rush of power that comes from spying on the lives of others. Some operatives-for-hire such as Christopher Steele are ex-government spooks or retired investigators with the FBI or other law enforcement agencies, looking to extend their careers by selling private clients the skills they had acquired while serving the public. The business also attracts former prosecutors and attorneys who aren't interested in working within the confines of traditional law firms. As newsroom jobs have vanished over the past decade, journalists such as Glenn Simpson and Peter Fritsch have joined the ranks of private operatives. Investigative firms also serve as homes to misfits, oddballs, also-rans, wannabes, and the occasional sociopath.

Getting a fix on the investigative industry's true financial size is impossible because it is sprawling and many firms within it are privately held. But one consulting firm that follows the business,

ERG Partners, estimated that its revenue in 2018 reached $2.5 billion, an amount representing a doubling of that figure from a decade earlier.

Opinions vary about the investigative industry's biggest players but such lists often include Navigant Consulting, FTI Consulting, Control Risks, Kroll Associates, K2 Intelligence, the Mintz Group, and Nardello & Company. Many of those companies offer similar services, such as investigative support for lawyers defending companies or individuals from criminal or regulatory actions and performing background, or due diligence, reviews. Various firms have tried to distinguish themselves from competitors by emphasizing specialized expertise into computer security, forensic accounting, or knowledge about Russia, Africa, or other parts of the world.

In deciding about how to focus my reporting for this book, Fusion GPS and Orbis Business Intelligence were two natural choices because of the outsized roles that Glenn Simpson, Peter Fritsch, and Christopher Steele played in recent political events. Given its involvement in the Harvey Weinstein case, Black Cube cried out for attention. I also wanted to include a traditional corporate intelligence firm and K2 Intelligence fit that criteria, particularly in light of its history. The firm is the second coming of Jules Kroll, the person often credited with creating the modern-day investigative industry when he started his eponymous firm, Kroll Associates, in 1972.

By a happy coincidence, Fusion GPS, Orbis Business Intelligence, Black Cube, and K2 Intelligence proved to be good choices for another reason. The four companies had all opened their doors for business around 2010 and so offered a way to look at the changes that have taken place within the investigative industry over the past decade.

It soon became clear to me that in writing about private operatives I would need to examine another profession as well: my own, journalism. Reporters get leads or tips from all kinds of sources. Some of them, such as consumers or government whistleblowers, want to alert the public to what they see as a danger or a wrong.

Journalists also have long gotten tips from hired operatives. But their relationship to private spies is different: it is a symbiotic and hidden one that benefits both sides. A reporter can obtain material from a private spy that they can't legally or ethically acquire elsewhere, things like stolen emails or confidential financial records. A private operative uses a reporter to make information public that benefits a client or damages an adversary without leaving their fingerprints behind. The journalist gets a scoop and typically never reveals how they acquired sensitive information. Everyone is happy and the public—the reader or the viewer—is left in the dark without any idea of what took place behind the curtains.

It's not surprising that the feverish embrace of Christopher Steele's dossier would unfold as the nature of the news business was changing. More media outlets were springing up on the internet, reporting had become more partisan and politicized, and the toxic interchange that is Twitter was in full bloom. Those changes would serve as a perfect petri dish, where the influence of the private spies would fester and breed, uncontrolled and unchecked.

CHAPTER 1

JOURNALISM FOR RENT

WASHINGTON, D.C., 2009

It probably sucked. There was no other way to put it. Glenn Simpson thought of himself as one of the top investigative reporters in the United States, a member of a swashbuckling elite he once described as "the Priesthood." But just as he was receiving public recognition of that status, he was pulling the plug on his career.

When the Trump dossier emerged in 2017, Simpson would achieve a new ambition—he would go from someone reporting the news to someone making it. But a decade earlier, he was fast approaching the end of his career as a reporter at *The Wall Street Journal*, a job he had cherished for fourteen years. He was invited in 2009 to speak alongside some of his profession's biggest names at a prestigious conference held at the Berkeley campus of the University of California. His fellow attendees included the then-editor of *The New York Times*, Bill Keller; the executive producer at the time of the celebrated PBS documentary series *Frontline*, David Fanning; and an investigative correspondent then at ABC News, Brian Ross. The organizer of the annual meeting, Lowell Bergman, was also a star in the world of investigative reporting. He had been a producer at *60 Minutes* during the glory days of the CBS News program, and had, among other scoops, gotten a former cigarette company executive to reveal on air how the industry manipulated nicotine. In *The Insider*, a 1999 movie made about the episode, Al Pacino played Bergman and Russell Crowe starred as the whistleblower.

The theme of the 2009 Berkeley conference, "Reporting on Corruption," was right in Simpson's wheelhouse. He was known as an aggressive reporter skilled at uncovering corruption in politics, business, and government. Journalism was a perfect profession for him, one that prized his talents, indulged his obsessions, and kept his excesses in check. Besides, his appearance and attitude toward authority likely never would have cut it outside a newsroom.

One ex-colleague dubbed him "Shaggy." It was an apt description, though a generous one. "Schlubby" would have worked, too. Simpson often arrived at work at the *Journal*'s Washington, D.C., office dressed in jeans, a threadbare button-down shirt, and a badly cut sports jacket that hung on his six-foot-three frame. He wore glasses, had a mop of dark hair, and sported a goatee. He had a mischievous grin and his skin had an unhealthy pallor, the apparent result of too much drinking, too little exercise, or both. There was an unnatural downward tilt to Simpson's chin and when he turned his head, he appeared stiff and slightly robotic. His movements were aftereffects of the spinal fusion Simpson had undergone while in college to repair a broken neck suffered in a car accident.

Simpson's desk was a thicket of documents and detritus, and he seemed to take a certain pleasure in scratching his belly or belching in public. Those habits aside, many of Simpson's fellow reporters liked him. He was warm, generous, funny, and a cutup who liked to pull pranks. The editors who managed him—or who tried to—had different feelings. They found him difficult, combative, and quick to see conspiracies where ones didn't exist. Some editors, upon learning that Simpson had just sent in a story for editing, would find an excuse to leave the newsroom in the hope that the story would get assigned to a colleague before they got back. One *Journal* editor became so concerned about the conclusionary leaps that Simpson was capable of making that he asked another journalist on the paper's staff to double-check Simpson's reporting. His response to any pushback he got from editors was usually the same: they could go fuck themselves, because if they knew how to report they wouldn't be sitting in an office on their asses. Along with talent, Simpson also didn't lack in self-confidence.

But the speech he was going to give at the 2009 Berkeley conference was one he never had expected to deliver. He was at the top of his game and generating the kinds of scoops that other reporters envied. Even so, he was going to use the occasion to announce that he was leaving journalism for a new career—he was going to become an operative-for-hire.

THE EVENTS LEADING TO Simpson's decision had started to unfold years earlier. By the mid-1990s, online news organizations were appearing and readers and advertisers began flocking to them. The newspaper industry started to collapse and papers nationwide soon were laying off hundreds of reporters and cutting back on expenses, including costly investigative projects that were Simpson's specialty.

The Wall Street Journal, with its loyal audience of business readers, was insulated from much of the chaos. But in 2007, it was shaken by a different kind of upheaval when News Corporation, the media conglomerate run by Rupert Murdoch, paid $5 billion to buy the *Journal's* parent company. The purchase set off alarm bells at the *Journal* because Fox News and other Murdoch-owned outlets covered the news with a slant that reflected his conservative politics. Murdoch promised to keep his hands off the paper's news-gathering operations but it didn't feel that way to many reporters there, particularly those like Simpson who were based in Washington, D.C.

After the 2008 election of President Barack Obama, *Journal* reporters felt whipsawed as editors in New York wrote headlines for their articles about Obama policies that they believed negatively slanted their reporting. Staffers also faced pressure to produce more articles, and Murdoch had made it clear in public comments that he didn't like the deeply reported articles and investigative exposés that long had been hallmarks of the *Journal*'s front page.

GLENN SIMPSON DESPISED MURDOCH. Still, the prospect of leaving journalism was wrenching. His identity had become intertwined with his job, and he had long seen himself as the type of reporter who was on a mission to expose wrongdoing and hold the powerful to account. His wife, Mary Jacoby, was a reporter, too, and so were many of his closest friends, who shared his obsession with political intrigue and corporate corruption.

Simpson, Jacoby, and their two sons lived in a large, ramshackle house near a Washington, D.C., neighborhood called Friendship Heights. Their home was the center of a lively social scene and the couple threw frequent parties attended by journalists, congressional staffers, and Washington insiders that were fueled by lots of alcohol and plenty of pot. On warm summer evenings, friends gathered at his house to drink wine, get stoned, play guitars, and listen to music.

Simpson was a voracious reader, especially biographies and works of history. He was also a fanatic about music and performers such as Alejandro Escovedo and the Drive-By Truckers, who fit his self-image as a kick-ass, rock-and-roll cowboy. After he became rich as a private operative, he liked to drive his expensive, tricked-out pickup truck to the second home he owned on the Maryland shore. For travel in Washington, D.C., he preferred another mode of transport, an electric scooter.

DIFFERENT REPORTERS HAVE DIFFERING skills. Some know how to get strangers to open up to them. Others see a story before competitors do. Simpson was a document hound. He loved finding obscure government and legal records that he then used to track transfers of money and hidden connections between people and companies. His first full-time job was at *Roll Call*, a Washington, D.C.–based publication that covered politics and legislation, and he specialized there in reporting about the financing of political campaigns. Candidates for federal office are required to make filings with the Federal Elections Commission that disclose their financial supporters and campaign expenditures and Simpson relentlessly mined those records for stories.

Simpson also met wife, Mary, at *Roll Call*. The couple shared journalistic interests and quirky personalities. Their appearances and backgrounds couldn't have been more different. Simpson towered over Jacoby, who was short, dark-haired, and stylish. She came from one of the wealthiest families in Little Rock, Arkansas, where her father was a top executive at Stephens Inc., a financial services firm that was a major political player in the state. They married in 1994 and, soon afterward, Jacoby's father helped the new couple buy their Washington home.

After the attacks of 9/11, Simpson made terrorist financing his beat and wrote articles examining how terrorist groups like Al-Qaeda were moving money to finance their activities by using loopholes in the international banking system or sham charities. Then his byline disappeared for a time. When he was in his late thirties, Simpson was diagnosed with spinal cancer, a painful and often-fatal form of the disease. Surgeons successfully treated the tumor but Simpson had to undergo treatment to help him get off the powerful opioids that doctors had prescribed for his pain.

IN 2005, THE FOUNDATION for Simpson's future career as an operative-for-hire was laid when the *Journal* made him a foreign correspondent based in Brussels. The paper also hired his wife, Mary, as a reporter and they moved with their children into a Brussels neighborhood favored by expatriates.

By then, Europe was awash in oligarchs, wealthy businessmen from Russia or Eastern Europe who had made fortunes in the years following the fall of the Soviet Union. The oligarchs liked to portray themselves as Western-style capitalists. Law enforcement authorities in the United States and elsewhere suspected that many of them had attained their power and wealth through political connections, bribery, violence, or criminal ties. The State Department also put some oligarchs on a shit list that barred them from entering the United States to conduct business.

For private spies, the emergence of the oligarchs produced a business bonanza, particularly for firms operating in London, a city that is an epicenter of the corporate investigations industry. Transplanted oligarchs bought up swaths of high-priced real estate there and sent their children to elite private schools. They listed their companies on the London Stock Exchange and used courts in England and elsewhere in Europe to engage in nonstop litigation against one other. In the process, they hired teams of private operatives to dig up dirt on their adversaries or intimidate their critics. One prominent oligarch, Oleg Deripaska, who was the kingpin of Russia's aluminum industry, used several U.S. or British investigative firms.

While based in Brussels, Simpson developed a network of private operatives who became sources for his next reporting obsessions: the financial and political worlds inhabited by oligarchs and Russian criminals. Business corruption in Russia was rampant under the reign of its president, Vladimir Putin, and Simpson's articles for the *Journal* about oligarchs were laced with allegations of bribes, money laundering, and the occasional murder. One of those articles started with the image of a Russian oil company executive, who after a hearty lunch of dumplings accompanied by shots of vodka, keeled over dead, the apparent victim of foul play.

SIMPSON ALSO CONNECTED WHILE in Brussels with people who would play significant roles in his later life as an operative-for-hire. Prime among them was an investor, Bill Browder, who became famous for his role in spearheading a lobbying campaign in the 2010s that led to adoption of a U.S. law that imposed financial sanctions on Russians involved in human rights violations.

Browder recalled he first met Simpson in 2006, shortly after the investor was expelled from Russia, where he had worked for a decade. While there, Browder had made a fortune operating Hermitage Capital, a Moscow-based fund, that held investment stakes in Russian companies worth $4 billion. During those years, Browder had praised Vladimir Putin's leadership but he earned the Russian leader's enmity after criticizing corruption in the country's energy industry and was barred from returning there.

In 2006, Simpson wrote an article for the *Journal* that mentioned Browder's expulsion. The investor would say years later that he and Simpson hit it off back then. Simpson shared the story of his recent battle with cancer, telling Browder how the ordeal had left him with an unexpected gift, a new sense of fearlessness. Simpson went to the London offices of Hermitage Capital, where the firm had relocated, to get information from Browder for articles he was researching about Russian business corruption. Neither man sensed how Simpson's transition into a private operative would turn them into blood enemies.

WHEN GLENN SIMPSON RETURNED with his family to Washington, D.C., after a few years in Brussels, he brought with him his interest in oligarchs and Russian organized crime. And in 2007, he and Mary Jacoby collaborated on articles for the *Journal* about two people who would remain fixations for him. One was Oleg Deripaska, the Russian aluminum oligarch and power user of corporate investigative firms. The other was a well-known Washington lobbyist, Paul Manafort, who worked in the mid-2000s to help a Putin-backed politician win the presidency of Ukraine, and re-emerged a decade later as Donald Trump's campaign manager.

Simpson's friends knew he had a tendency toward paranoia and upon his return to Washington, it found a new outlet. He worried that Russian spies or operatives working for oligarchs were bugging his home. Meanwhile, his intoxication with the cloak-and-dagger world of private spying and its intrigue was growing. "He loved the skullduggery and the amoral nature of the business and liked to say, 'My guy told me this,'" one *Journal* reporter recalled.

NEAR THE END OF his time at *The Wall Street Journal*, Simpson would write a series of articles that highlighted his reliance on hired operatives for tips and foreshadowed what lay ahead for him. The stories, which were published in 2008, were about a feud unfolding in a place few Americans had heard about: Kazakhstan, a former Soviet republic in Central Asia.

At the time, the country was best known to outsiders through *Borat*, the mock documentary in which the British comedian Sacha Baron Cohen portrays a fictional Kazakh journalist who travels to the United States in search of Pamela Anderson, the star of the television show *Baywatch*. But the oil-rich country was also home to a massive and long-running bribery scandal that started in 2003 when the U.S. Justice Department charged an American consultant with giving $80 million in payoffs to Kazakhstan's president and others to win drilling rights there for U.S. energy companies.

When Simpson started writing about Kazakhstan in 2008, the country's ruler, Nursultan Nazarbayev, was pitted in a political death match against his former son-in-law. Both men were horrible. Nazarbayev was a ruthless dictator whose enemies and critics disappeared. His onetime son-in-law, Rakhat Aliyev, was a wannabe dictator who reportedly tortured people while he was head of his country's secret police. The two men fell out after Nazarbayev announced plans to be president for life and Aliyev, who had political ambitions, criticized the move. In a heartbeat, he was stripped of his official positions, his wife divorced him, and a Kazakh court convicted him in absentia of various crimes, including kidnapping and murder.

Readers of *The Wall Street Journal* likely had limited interest in this internecine feud but they were soon treated to a stream of articles about it. Both the Kazakh government and Rakhat Aliyev had hired high-priced American lawyers, political operatives, and corporate investigators and Aliyev's team hoped to gain political asylum for him in the United States. They launched a campaign portraying him as a Western-style politician eager to free his homeland from Nazarbayev's corrupt yoke, and, as fate would have it, Simpson knew players on Aliyev's team.

Aliyev hired a big U.S. law firm, BakerHostetler, to lobby on his behalf, and Simpson had ties to attorneys there. His connections to one of them, John Moscow, dated back to the 1990s, when Moscow was a top prosecutor in the Manhattan District Attorney's Office. Simpson was a reporter at the time for *Roll Call* and covered Moscow's cases, including his prosecution of the Bank of Credit and Commerce International, a notorious financial institution that laundered money for drug dealers and others.

GLENN SIMPSON WAS SOON getting the type of material that a reporter usually only dreams about. Those documents—personal emails, confidential bank statements, credit card statements, cellphone records, etc.—detailed the personal and financial dealings of a consultant in Washington, Alexander Mirtchev, who was a top advisor to Kazakhstan's sovereign wealth fund and a main target of Rakhat Aliyev.

In mid-2008, Simpson and another investigative reporter at the *Journal*, Sue Schmidt, began writing a series of articles about Mirtchev and the Kazakh bribery scandal that depicted the consultant as a political and financial fixer for Kazakhstan's president. Schmidt was a recent arrival at the paper, having come there just a few months earlier from *The Washington Post*, where she and two colleagues had won a Pulitzer Prize in 2006 for exposing a corrupt lobbyist, Jack Abramoff. Her tenure at the *Journal* would be extremely brief. About a year after joining the paper, she left with Simpson to form a corporate investigations firm.

FOR ONE FRONT-PAGE ARTICLE, they interviewed Rakhat Aliyev, who was in hiding. Aliyev claimed that his former father-in-law had stolen billions of dollars from Kazakhstan and that Alexander Mirtchev had helped him hide the funds. He gave the reporters what he said were bank statements and wire transfers from Mirtchev's companies.

The article reported that Mirtchev was the subject of criminal investigations. "The Federal Bureau of Investigation and the Manhattan district attorney are looking into Mr. Mirtchev's banking activities, according to lawyers and businessmen who have had contacts with investigators," Simpson and Schmidt wrote. "The inquiries are said to be at an early stage. Representatives for the Justice Department and the district attorney's office had no comment."

Mirtchev denied he had done anything wrong, and he and his lawyers insisted that Aliyev was handing out forged or altered documents. Mirtchev also tried to make his case to the reporters but those meetings didn't go particularly well. Journalists are never happy when their reporting is questioned, but Simpson's meetings with Mirtchev turned confrontational. The same traits that made him a talented reporter—his relentlessness, his self-assuredness—had a flip side that emerged when he was challenged. His response to criticism was to double down. His self-confidence could flare into arrogance.

At one meeting with Mirtchev, Simpson opened his laptop and displayed to the consultant all the confidential information he had obtained about him, according to an account of the meeting written by a public relations executive who accompanied Mirtchev to it. Those documents included copies of Mirtchev's personal and business emails, his credit card statements, and corporate bank records.

Simpson told Mirtchev that he personally knew how to get nonpublic information and, if he couldn't get it himself, he knew people who could get it for him, according to the public relations executive's account. In addition, Simpson said that he and Schmidt were able to corroborate their claims because some of Mirtchev's telephone conversations had been wiretapped. Mirtchev reportedly expressed outrage, accusing Simpson of invading his privacy, and he insisted that Rakhat Aliyev was distributing fake documents. Simpson was unimpressed and apparently told Mirtchev at one point that he knew more about him than Mirtchev knew about himself, the public relations executive's account stated.

AS A JOURNALIST, GLENN Simpson couldn't hack into somebody's email account or gain access to bank computers. He couldn't pay an operative or a corporate insider to steal documents for him. But like any reporter, he was free to accept stolen or hacked information from anyone who wanted to give it to him and he wasn't under any obligation to disclose—other than possibly to his editors—where he had gotten the material.

But back in 2007, well before Simpson took an interest in Alexander Mirtchev, a massive private spying and hacking campaign was launched against the consultant. And that operation, apparently underwritten by Rakhat Aliyev, would harvest precisely the same type of financial and business information that Simpson would display a year later to Mirtchev during their meeting.

That operation, dubbed "Hellenic," unfolded at a pivotal moment in the evolution of the private intelligence industry, one where hired operatives began using hacking, email surveillance, and other high-tech tactics to secretly garner material. The private spying firm that conducted the investigation didn't identify itself in written reports it produced about the probe. But the operatives involved in Operation Hellenic were eager from the start to get the information they collected into the hands of journalists and law enforcement officials. According to one 2007 memo, an objective of the spying campaign was to bring negative information about Mirtchev to "the attention of the press and investigative authorities," including material that could be used to bring money-laundering charges against him.

The private spies involved in Operation Hellenic employed a variety of tricks, according to their reports. A female operative posing as a journalist was dispatched to interview Mirtchev's business partner and a type of malware known as a "trojan" was deployed to infect a computer that Mirtchev and his partner used. "An intercept package has been delivered via intermediaries in the US which if installed in his computer with internet access should allow us complete monitoring capability," a 2007 report stated.

That electronic surveillance plan seemingly succeeded. In a subsequent report, investigators wrote that they had intercepted six hundred emails written or received by Mirtchev and his partner and were in the process of examining four hundred more. In describing their capabilities to remotely monitor the consultant through his computer, operatives wrote:

- We see every email he sends with attachments
- We see every website he visits
- We see the username and password he uses
- We see what he types into documents and spreadsheets
- We see the password or passphrases he uses to secure these files

- We see the password or passphrases used to encrypt or decrypt documents
- If he uses chat rooms we see everything he says
- We get email attachments if he sends or receives them

By early 2008, Operation Hellenic investigators stated in a report that they were ready to "present a money laundering case to US Federal authorities" about Mirtchev. Four months later, Simpson and Sue Schmidt started writing about the consultant and reported that the FBI and the Manhattan District Attorney's Office had opened inquiries into him.

FOR A JOURNALIST, DECIDING when or even whether to report about the existence of a potential criminal investigation is a tricky business. Prosecutors are barred from disclosing that such inquiries are under way until charges are brought, so reporters typically learn about ongoing investigations from lawyers or other people connected to a case. The trouble is those sources come with an agenda. Law enforcement investigations often start after prosecutors are approached by lawyers for companies or the wealthy who believe a crime has taken place. What that means is that the two-tiered system of justice in the United States isn't simply one in which the poor go to prison and the rich face lighter punishment, if any, for similar crimes. It is also one in which the powerful aided by their lawyers and private operatives have a substantial impact on the cases authorities decide to prosecute. When authorities don't respond quickly, lawyers or their clients can turn to another tactic. They find a friendly reporter and plant a story about a budding investigation as a way of trying to pressure prosecutors into action. Journalists refer to the technique as "front-running," or getting out ahead of a story's facts.

In the case of Alexander Mirtchev, Simpson took the plunge.

His articles didn't identify the "lawyers and businessmen" whom he cited as having spoken to investigators with the FBI and the Manhattan District Attorney's Office. But in a 2009 State Department cable, a U.S. diplomat reported that Kazakh government officials had complained to him that "materials, concerning Kazakhstan's leadership have been sent by Rakhat Aliyev with the help of Baker and Hostetler LLC to various U.S. law enforcement agencies—in particular, to the District Attorney of New York Robert Morgenthau and to the FBI."

Some journalists don't report about the existence of "preliminary" investigations for another reason. The term can mean all kinds of different things and early-stage inquiries may never progress. That's seemingly what happened in the case of Alexander Mirtchev. More than a decade after Simpson's articles were published in the *Journal,* a lawyer for the consultant would say that Mirtchev was never contacted by the FBI or the Manhattan District Attorney's Office about his financial dealings back in 2008 or at any point afterward.

BY 2009, GLENN SIMPSON's interest in Kazakhstan was over and his attention had shifted to a more pressing matter—his future. The exodus of talent from the *Journal* that started with Rupert Murdoch's acquisition of the paper was intensifying. Opportunities at *The New York Times, The Washington Post,* and other major papers were limited because those publications were cutting staff and some reporters were leaving journalism. In late 2008, Mary Jacoby, Simpson's wife, quit the *Journal* to start a new business, a website called Main Justice that covered the Justice Department and major cases.

Simpson and Sue Schmidt had started discussing a new venture of their own, one that would make use of their investigative skills and the connections they had made over their careers. They saw a void in the corporate intelligence industry and an opportunity to fill it with a different type of firm—one that embraced the values and ethical standards of journalism while working for private clients.

They would take assignments only from "good guys" such as public interest groups, nonprofit organizations, or companies with legitimate legal gripes. To market their new firm, SNS Global, they would tell potential clients how their journalistic reputations and ties to reporters would enable them to get publicity for their cases in major media outlets.

Over the years, Simpson and Schmidt had gotten to know plenty of private eyes and hired operatives. Some were likable. Others lied, worked for bad actors or criminals, or took on assignments to smear journalists. Neither of them wanted to think of themselves as private investigators. Instead, Simpson called the concept for his new career "Journalism for Rent."

Managers at *The Wall Street Journal* urged both of them to stay. But in April 2009, Simpson walked into the paper's office in Washington for the last time as an employee. According to a published account, he was reportedly accompanied by his dog, a dachshund-beagle mix named Irving, whom he encouraged to defecate by the desks of editors. It wasn't clear if he was joking.

SOON AFTERWARD, SIMPSON WAS in Berkeley for the investigative reporting conference. As his panel started, he sat slumped in a director's chair looking at the name tag around his neck. It identified him as a *Journal* reporter.

"I am actually leaving journalism," Simpson told the group. "Yesterday was my last day at *The Wall Street Journal* so don't accuse me of false advertising. I didn't have a chance to update the name tag here but I did just want to mention really quickly that we were going to try something new that is not going to be journalism per se, but is going to be sort of a hybrid.

"We are hoping to do this kind of investigative work and work with media organizations to bring these stories out but myself and Sue Schmidt, my partner, have formed a private company to see if we cannot pioneer as yet another new model to keep investigations going, keep doing this in the public interest," he added.

With that, Glenn Simpson's first career was over and his second one was starting. His friends in the investigative industry had given him some advice. He needed to stop thinking of himself as a journalist. He needed to avoid getting emotionally invested in cases and he needed to make sure that he got a deposit from clients before he started working for them. But most of all he needed to go out and buy some decent suits.

CHAPTER 2

"LAPDANCE ISLAND"

LONDON, 2012

When a freelance private spy named Rob Moore arrived in 2012 at the London offices of K2 Intelligence he was eager to hear about a new assignment. Moore, like Glenn Simpson, recently had embarked on a second career as a corporate investigator, though he was a refugee from a different industry—the entertainment business.

Moore was once a successful producer of television "prank" shows, comedy programs where the unsuspecting become the butt of a joke that the audience is in on. In the early 2000s, Moore was an executive at a well-known British production company, Ealing Studios, where he worked on a prank series that tricked people into auditioning for the pilot episodes of supposed reality-style television programs that would never be made.

One was called "Who Wants to Be Prince William's Girl-friend?" It was filmed in Los Angeles and young women who thought they were competing for a royal date had to jump through various hoops before being escorted to a throne and crowned. A phone then would ring and Moore, pretending to be Prince

William, was on the line. "Really liked your audition, I thought it was really great," he would say. "When you come over, could you bring two hundred fags?" he added, employing the British slang term for cigarettes. The high point of the series was a show modeled after the hit program *Survivor*. It was called "Lapdance Island" and the program's motto was "Forty Lap Dancers, Ten Contestants, Sun, Sand, and some Huts." Hundreds of men turned up in response to ads announcing auditions.

Sadly, the fun didn't last. In 2006, when Moore was in his early forties, his marriage and career collapsed and he found himself living back at home with his parents. He decided to pursue a simpler life. He embraced Buddhism and studied horticulture with an eye toward becoming a gardener. Gardening didn't last long, either. While walking on a beach, he ran into a friend from his television days who told him he was now working for Kroll, the big corporate intelligence agency. The firm, his friend said, would value Moore's talents.

CORPORATE INVESTIGATIVE FIRMS, LIKE most businesses, have hierarchies. There are the owners and bigger firms often have a board of directors or advisors, typically retired officials from intelligence and law enforcement agencies chosen for their rain-making abilities. Senior investigators who oversee cases and interact with clients are often called managing directors to lend them a corporate air and regular investigators work under them. (People in the investigations industry don't like the term "operative" because they feel it has negative connotations.) On the next rung down are researchers, who are known as analysts, and they dig out information from public filings, databases, and other records. At the bottom of the heap are freelance spies, or so-called contractors (who are also sometimes referred to as subcontractors or subs).

Some firms hire contractors on a temporary basis for specific assignments. Other firms won't use contractors to do sensitive investigative work because they can't control them. But in London's frenzied market for private spying, K2 Intelligence bolstered its ranks by hiring Rob Moore. Besides, he had a special skill.

MOORE HAD A NATURAL talent for deception, one that was an outgrowth of his television career. Real spies and private operatives have long pretended to be someone they are not—a cop, a bank officer, an employer, a distant relative—in order to con a stranger into giving them confidential information.

The technique is known as "pretexting" and Ian Withers, the longtime British private detective, considered himself a craftsman of the trade. Ian hired people to work at his detective agency who were ex-employees of banks, loan firms, or telephone companies and, as a result, knew the types of questions that a customer service representative would ask a caller to verify their identity and the best ways to fool them. Ian also believed in verité when it came to fostering illusions, so people making pretext calls from his agency worked out of a special room equipped with tape recorders that played background noise suitable to the identities they had assumed that day. For instance, when they were pretending to be loan officers, background sounds made it seem like they were calling from a bank.

Early in his career, Ian's reputation for stealth became so widespread that he got an unexpected call in 1971 from *The Guardian* newspaper in London with an invitation to join its editor for lunch at his private gentlemen's club. As they dined, the editor explained he had made a bet with England's prime minister, Edward Heath, who had claimed during a speech in Parliament that confidential information collected by his government about British citizens could only be accessed by those authorized to do so.

The editor gave Ian the names of four employees at the newspaper, including his own, and told him to have fun. Not long afterward, a front-page article was published in *The Guardian* under the headline "Commercial Spies Tap State Records." Heath was not amused. The police started raiding detective agencies in London. At the one run by Ian and his brother, they opened a vault and found hundreds of reports crammed full of private financial or medical information about people his firm had investigated. Ian and his brother were charged with conspiracy to commit a public mischief. At his trial, a prosecutor asked Ian whether he used pretexting to obtain confidential information. "What do you think?" he replied. He and his brother were sentenced to a year in prison but their convictions were later thrown out.

ROB MOORE COULD EASILY pretend to be someone else when speaking to a stranger over the phone. But he also could do it face-to-face with someone he had never met. Moore was tall and had a high forehead and a long, narrow face with plastic features that easily shifted between expressions of interest, concern, and sympathy, all as a tape recorder hidden in his jacket or shoulder bag captured a conversation.

One of the first freelance assignments Moore got from Kroll was to find out who owned a mansion in London where an oligarch was living. To do the job, the ex-producer decided to pose as a deliveryman from a high-end floral shop. He went to a housewares store, bought a gardener's smock, and then picked up a large bouquet of flowers. When he checked himself out in a store mirror, he was so amused with his appearance he took a selfie. Then he arrived at the mansion and talked his way inside.

PRETEXTING STILL LIES AT the heart of the private spying business. For instance, operatives for Black Cube, the Israeli firm that worked for Harvey Weinstein, have adopted all kinds of poses to lure targets into compromising situations. On various assignments, they pretended to be business executives, job recruiters, investment bankers, concerned mothers, or advocates for various causes. But private spies really like to pretend they are journalists.

That makes sense because people expect a reporter to be nosy and ask questions. And to effect that ruse, investigative firms have a couple of options. They can hire a freelance reporter and have him or her pretend to be gathering information for an article when what they really are doing is spying. At other times, a private investigator will cut out the intermediary and pose as a reporter.

One journalist wrote about how a Kroll executive approached her to go undercover for the firm. The reporter, Mary Cuddehe, described how the Kroll official wanted her to travel to the Ecuadorian Amazon, and present herself as a journalist doing an article about an environmental dispute but actually collect information for a Kroll client. Cuddehe was a struggling freelancer and she was offered a lot of money: $20,000 for six weeks of work. She was tempted but decided instead to write an article for *The Atlantic* in which she outed Kroll's attempt to recruit her. At the time of the incident, Jules Kroll was no longer involved with Kroll the company. "At first, I thought I was under qualified for the job," Cuddehe wrote. "As it turned out I was exactly what they were looking for: a pawn." After several embarrassing episodes, Kroll stated publicly that their operatives are not supposed to work in this way and said it was instituting new ethical guidelines to govern those practices.

AT THE START OF his career as a private operative, Rob Moore's assignments were short term and sporadic and he continued to struggle to make a living. But that all changed in 2012 when he was buzzed into the offices of K2 Intelligence inside a red, sandstone building at the edge of Mayfair, a high-end London neighborhood that is home to corporate investigations firms and the fashionable stores of Bond Street and Savile Row.

A receptionist escorted Moore down a hallway lined with small meeting rooms that had frosted glass windows. He could only see the tops of the heads and the shoes of those seated inside. He was ushered into an empty room and soon joined by an executive of K2 Intelligence who was the head of its London office. That executive, Matteo Bigazzi, told Moore that a new client wanted the firm to investigate a group of public health activists campaigning to end the use of asbestos, a building material linked to a deadly type of lung cancer. To infiltrate the group, Moore would pretend to be a maker of documentary films that exposed health threats.

K2 INTELLIGENCE WASN'T REALLY a new firm when it opened up in 2009. Instead, the K2 in the firm's title stood for Jules Kroll, a creator of the modern-day corporate investigations industry, and one of his sons, Jeremy Kroll, who had followed his father into the business.

Five decades earlier, when Jules Kroll started his first company, Kroll Associates, the public image of a private investigator was that of a back-alley snoop who spied on cheating spouses. In the simplest terms, Kroll changed all that by corporatizing the business of private spying. His overarching goal was to give Kroll Associates the same veneer of respectability that law firms enjoyed. It was a strategy that also allowed him to charge similar fees. One former employee of K2 Intelligence said that when he asked Jules Kroll how he should provide clients with an estimate of a case's cost, Kroll offered this simple formula: "He said, you should figure out what you want and then triple it," that ex-employee said.

To rebrand the investigations business, Jules Kroll reached out beyond the usual cadre of ex-spooks and retired agents and cops who staffed most firms and hired lawyers, forensic accountants, former journalists, and recent college graduates who were bright, curious, and presentable. The approach reflected his own unplanned career path. He had first worked as a prosecutor in the Manhattan District Attorney's Office and dreamed about a career in politics. That hope died after he was soundly defeated in his first attempt to win elective office and he never ran again. Instead, he decided to take a flyer on an idea that he said had come to him when he took over his family's commercial printing business for a time while his father was ill, according to one account of his career.

Back then, printers paid kickbacks to brokers who controlled printing jobs, bribes that inflated costs for publishers, advertising agencies, and other companies that spent heavily on printing. Kroll realized that he could make money by helping businesses save money by ending the practice. In recounting his origin story to an audience of college students, Kroll explained that his big break came when a major publisher of comic books, Marvel Comics, hired him. "Spider-Man, the Hulk—I really owe my business to those superheroes," he said, according to a profile in *The New Yorker*.

In the 1980s, Kroll Associates was perfectly positioned to take advantage of the corporate takeover boom. One major Wall Street bank, Drexel Burnham Lambert, hired the firm to conduct due diligence investigations on potential investment targets to see if those companies had hidden problems. To gather material, Kroll investigators and researchers scoured lawsuits, property deeds, corporate filings, and other kinds of public records. They also collected trash outside a target's home and shadowed people, all of which is legal.

Fortune 500 companies facing hostile takeover bids also

turned to Kroll Associates to unearth embarrassing information that would repel them. One well-known corporate raider of the 1980s, T. Boone Pickens, complained that Kroll Associates had a two-foot-thick dossier on him that it offered to sell for $500,000 to any company he tried to take over.

Soon companies were paying anywhere from $15,000 to $50,000 to Kroll Associates and other big corporate intelligence firms to do "due diligence" or background investigations of potential business partners, takeover targets, or investments. Initially, those reviews were only run on businesses or people with suspected baggage but in time it became standard business practice for companies to hire Kroll or another firm to do them as a hedge against potential legal liabilities if a deal or a business partnership went south. Law firms increasingly hired Kroll or one of its competitors to help prepare cases for litigation or investigate an opponent. The going daily rate for each operative was $1,000 a day, plus expenses.

The New York Times dubbed Jules Kroll "Wall Street's private eye," a title relished by the investigator, who was tall, athletically built, and liked to wear braces. By the late 1980s his firm had grown to more than two hundred employees and had offices in New York, London, Washington, Los Angeles, and Hong Kong. Working as a corporate investigator became very lucrative and top executives at big firms such as Kroll made up to $300,000 annually.

The job of private operative also acquired panache. When the San Francisco office of Kroll Associates announced openings for two entry-level positions, it received more than three hundred applications from college students. "Many of us are corporate misfits," one attorney told a reporter, explaining why she had left a big law firm to join Kroll. "These are people who are more comfortable asking questions and think that making trouble is a good thing. They all think it is great fun."

At one time, the client roster of Kroll Associates included a future U.S. president, Donald J. Trump. The real estate developer hired the firm in the mid-1980s to find out if organized crime figures were involved in an Atlantic City casino he wanted to buy. Trump became furious, Jules Kroll told a reporter, when he learned that the firm's investigators had given the casino a clean bill of health. The developer wanted the report changed, Kroll remembered, because he believed that having gangsters involved in the casino—or at least having a report that said they were—would drive down the casino's value and increase his negotiating leverage. "He wanted us to rewrite the report," said Kroll, adding he refused to do so and fired Trump as a client. When told at the time about Kroll's comments, the future president responded in a typical fashion. "That's crazy," he said. "Whoever heard of something like that!"

JULES KROLL BLAZED ANOTHER path that other hired operatives followed—he knew how to work the news media. Public relations firms typically acted as intermediaries who provided journalists with material that private investigators had gathered. But during the financial takeover boom of the 1980s, Jules Kroll developed a coterie of reporters to whom he regularly fed tips.

One of his biggest marketing coups came in 1991 when *60 Minutes* broadcast an episode about how his firm had been hired to track down the billions of dollars that Saddam Hussein, Iraq's leader, had skimmed off from his country's oil sales and hidden in banks or companies worldwide. "He has put together a network through some very clever colleagues that is extensive and far-reaching and pernicious," Kroll said.

Kroll Associates and *60 Minutes* did uncover key information about where Saddam had hidden his stolen assets. But an article in

New York magazine that was published soon after the show aired pointed out that Kroll was claiming credit for discovering information about the Iraqi dictator's holdings that others had already brought to light.

BY THE MID-1990S, THE industry that Jules Kroll had helped create was changing. Big accounting firms, seeing the profits that could be made, opened investigative divisions, and private operatives who had cut their teeth at Kroll Associates founded their own firms. Corporate intelligence companies promoted themselves using similar lingo. They offered "strategic intelligence," "risk consulting," "dispute resolution," and other catchphrases. More significantly, intensifying competition had turned due diligence investigations and other common services into commodities, cutting the prices firms could charge. As computerization increased, some investigative companies offered bulk rates to screen the histories of potential employees, charging $10 to $50 a head.

There was plenty of traditional investigative work still available but to keep profits high, private spies had options. They could work for oligarchs or other unsavory clients and adopt the more aggressive tactics those customers preferred. There were also new targets, including activists, nonprofit organizations, politicians, and members of the news media.

ONE SIGN OF THE industry's new direction emerged in 2001 and involved a high-profile London-based investigations firm, Hakluyt, which was staffed by former MI6 operatives. That year, more than a decade before K2 Intelligence dispatched Rob Moore to pose as a documentary filmmaker, Hakluyt used a similar strategy. It deployed an operative to infiltrate Greenpeace, the environmental group, under the guise of being a crusading filmmaker.

At the time, Greenpeace was protesting oil drilling projects in Nigeria by two energy giants, Shell and BP, and the firms hired Hakluyt to find out about the group's plans. In Hakluyt's case, its operative had a somewhat different background from Moore. He was a German-born spook with ties to that country's intelligence agency who had earlier infiltrated terrorist groups. Along with Greenpeace, Hakluyt reportedly used him to spy on the Body Shop, the natural cosmetics company, which was running a campaign in the early 2000s opposing Shell's drilling in Nigeria.

When the episode came to light, British lawmakers called for an investigation into ties between spies-for-hire and government spying agencies. But the inquiry never happened for a simple reason: private spies and government spies share information all the time and no one was about to talk about it.

Not long afterward, another major investigative firm, Diligence, got caught after it had pulled off an audacious caper. Diligence was headed by former staffers at U.S. and British intelligence agencies, and it was hired in the mid-2000s by a major Moscow-based financial institution, Alfa Bank. At the time, the bank was battling an investment fund located in Bermuda for control of a Russian telecommunications company.

KPMG, a big accounting firm, was conducting an audit of Alfa Bank's rival and the Russian bank hired Diligence to get an early copy of the report before it became public. To do so, Diligence launched an operation worthy of a spy novel, according to an account in *Bloomberg Businessweek* magazine.

Diligence's English cofounder, Nick Day, had a background in espionage. He was an alumnus of MI5, Britain's domestic intelligence service, and he approached a KPMG employee in Bermuda telling him he was a secret agent dispatched by Her Majesty's government to obtain the audit because it contained national security information.

The KPMG employee wasn't selected at random. Instead, Diligence developed psychology profiles detailing the type of person most likely to leak information. In the case of a man, he would have "a propensity to party hard, needs cash, enjoys risks, likes sports, likes women, is disrespectful of his managers, fiddles his expenses but is patriotic."

Once targeted, Day drew the KPMG employee into his trap by making him feel like he was taking part in a real-life spy thriller. He was instructed to go to public parks in Bermuda and place documents under marked rocks, a technique used by spies called a "dead letter drop." Day wrote in a memo, "We are doing it in a way that gives plausible deniability, and therefore virtually, no chance of discovery."

Diligence got the audit and Day gave the duped KPMG employee an expensive watch. On the back of it, a jeweler had engraved a supposed note of gratitude from the British government. But Day had apparently pissed off a former Diligence colleague and his firm ended up getting sued by the investment fund audited by KPMG after memos about the operation were leaked.

AROUND THAT SAME TIME, a major private spying scandal involving a big American computer maker, Hewlett-Packard, was unfolding in the United States. *The New York Times*, *The Wall Street Journal*, and *CNET*, an electronics industry publication, began reporting on discussions taking place inside HP's boardroom. Company executives wanted to find out which board members were leaking to journalists and hired a private investigator.

What happened next was indicative of how the private spying industry handles its dirty work. Jobs get farmed out through chains of contractors and subcontractors in an effort to insulate the beneficiary of material that was hacked or obtained through sleazy means.

In the HP case, the job of acquiring the phone records of board members and reporters was given to an "information broker," a breed of operatives flourishing in the mid-2000s. Telephone companies, banks, and credit card issuers, after getting stung by pretexters for years, had adopted stronger security measures to protect customer information. But information brokers—really pretexters under a different name—were using new commercial databases to vault over those safeguards.

The databases contained vast reservoirs of information about individuals, including their birth date, home address, and, sometimes, Social Security number. And armed with those details, information brokers were able to get past improved security procedures and do a thriving trade selling their services to investigative firms, lawyers, and others eager for information on targets.

The HP episode emerged publicly when an irate board member learned about the spying and congressional hearings examining the practices of information brokers followed. In time, a federal law was passed that made it illegal to use deception to obtain information from a financial institution about a customer. But a new and improved technology would soon eliminate a need for any interpersonal interactions—the era of cyber-spying would begin.

AROUND THE TIME OF the HP scandal, Jules Kroll's well-crafted reputation also started to spring holes. Once a lawsuit has started, private investigators aren't supposed to misrepresent themselves to witnesses. But according to an account in *The Wall Street Journal*, Kroll operatives repeatedly did so in the 1990s while claiming they were government employees, book researchers, or consultants.

A law firm fired Kroll Associates after one of those incidents came to light. Kroll officials described such episodes as unwitting mistakes and said the firm was opposed to the practice. More issues were to follow. In 2004, police officials in Brazil raided the offices of Kroll's firm there and arrested several employees, charging them with wiretapping, bribery, and hacking. (Charges against most of those arrested were dropped.) For his part, Jules Kroll insisted his staffers had done nothing wrong and that the company his firm was investigating had bribed retired Brazilian police officials to conduct the raid.

With rare exceptions, few journalists subjected Jules Kroll's firm or its competitors to serious scrutiny. The reason was simple. Reporters knew that once they take on a corporate intelligence firm they could kiss it off as a future source for stories. The effect of this buddy-bonding was blindness.

For years, executives of Kroll (which changed its name to Kroll Inc. in 2001) liked to boast to reporters about what they called their "Hall of Slime," a kind of imaginary rogues' gallery of the worst actors the firm had investigated. As it turned out, Jules Kroll's company had counted one of the slimiest operators of the 2000s among its clientele. He was a financial con man and money launderer named R. Allen Stanford, who ran a massive Ponzi scheme that collapsed in 2009, costing investors hundreds of millions of dollars.

The role played by Kroll Inc. in protecting Stanford from scrutiny would emerge in the scandal's aftermath. One former FBI agent told *Vanity Fair* magazine that Kroll operatives acted as front men for Stanford, defending his reputation while law enforcement officials were trying to learn if his bank was laundering money. "Kroll [Associates] was essentially running a propaganda campaign in defense of Stanford's good name," that ex-FBI agent said. "They beat on me many times: 'Hey you got this all wrong, he's not a money launderer, he's a great guy, leave him alone.'"

Kroll operatives, acting on Stanford's orders, also attacked those seeking to expose him. In one case, the businessman instructed a Kroll operative to go after a U.S. Senate committee staffer and lawyer who Stanford believed was the source of a negative magazine article about him. He "is a pure cockroach," Stanford wrote about the lawyer, Jonathan Winer. "Go after him hard on as many fronts as possible." Reporters disclosed that the Kroll executive began chasing rumors that Winer's ex-wife had left him for another woman, a rumor Winer has called absurd.

THE FIRM'S DEALING WITH R. Allen Stanford highlighted another troubling practice within the private spying industry related to due diligence reviews. Originally, a client hired corporate investigators to investigate a company or an individual. But by the 2000s, oligarchs and other wealthy business operators were getting ahead of the game by paying Kroll Inc. and other firms to perform reviews on them, that they then handed out as evidence of a clean bill of health.

By their nature, self-due diligence reviews are riddled with all kinds of problems. A client might limit an investigative firm's inspection to selected parts of its operations. It is also a potential minefield for conflicts of interest and Kroll Inc., while working for R. Allen Stanford, apparently blew straight past one of the investigative business's few red lines: the unwritten rule that a firm has an obligation to notify a client if an assignment poses a conflict of interest.

A construction industry trade group that wanted to invest money in Stanford's bank hired Kroll Inc. to conduct a due diligence review of it. A Kroll executive allegedly never informed the group that Stanford was a client and sent it a glowing report attesting to his bank's financial soundness. Two months later, Stanford's Ponzi scheme collapsed, wiping out the group's $2.5 million investment. It subsequently sued Kroll Inc., which denied liability, and the case was settled before trial. For his part, Jules Kroll later described the Stanford episode as "clearly a blemish."

BY THE LATE 2000s, Jules Kroll was a millionaire many times over. He had once hoped to turn his investigative firm into a diversified corporation that would offer consulting, investment banking, and other services. Between 1997 and 2008, he became involved in a series of business machinations. He merged Kroll with an armored car maker and brought the combined companies public, then ditched the car company and acquired a series of firms that offered various services. Then, in 2004, he sold Kroll to a big insurance company for a whopping $1.9 billion. Four years later, in 2008, the insurer, realizing it had overpaid, put Kroll up for sale. Jules Kroll retired from his namesake firm and launched an effort to buy it back. But he got outbid and the investigative firm called Kroll, while retaining his name, would no longer employ him or any other member of his family.

K2 INTELLIGENCE, THE COMPANY founded by Jules Kroll and his son Jeremy, marketed itself as a top-tier corporate intelligence firm in the tradition of Kroll. Its motto reflected what the firm described as its professional and ethical philosophy: "Do no harm, do the right thing, and do what you promised to do. That's the K2 difference."

Jeremy Kroll, who was about forty when K2 Intelligence was founded, was bright and well intentioned. But he struck people who worked with him as a kind of lost soul who lacked his father's investigative instincts and had gone into the family business because he didn't know what else to do. His brother, Nick Kroll, was a well-known successful comedian, and Jeremy had hoped for a career in art. He studied painting in college and the New York offices of K2 Intelligence were decorated with his work. "He told me that he wanted K2 Intelligence to be like an eclectic artist colony of creative types," one former employee recalled. One of Jeremy Kroll's paintings hung on his office wall. It was a large abstract portrait of a person whose face looked incomplete or unfinished.

DURING ITS FIRST YEARS of operation, K2 Intelligence's profit center wasn't in New York, where the Krolls were based, but in London. On paper, the executive who met with Rob Moore to discuss the asbestos case, Matteo Bigazzi, was its head. But the driving force behind the firm's business there was an operative named Charlie Carr.

He had roguish good looks and wore expensive clothing with the casual ease of a British private school graduate. Carr had worked for Jules Kroll for decades and had a reputation as a lady's man. He also was known for pushing boundaries. He worked at Kroll's offices in Brazil around the time it was raided by the police. He then went to the firm's office in Milan, which Matteo Bigazzi was then running. He met the sister of Bigazzi's wife and married her.

After Bigazzi left Kroll Inc. to open K2 Intelligence's London office, Carr joined him. Several years later, Carr made headlines when he stood up in a London courtroom, whipped out a camera from underneath his coat, and took a photograph of a witness while he was testifying, in a bid to unnerve him. Under British law, Carr could have been sent to prison but somehow dodged the bullet.

In the early 2010s, K2 Intelligence's profits in London were coming from clients who were oligarchs or multinational companies with operations in Africa or other developing regions. Carr worked closely with those free-spending oligarchs or their lawyers and the Russian aluminum magnate, Oleg Deripaska, was reportedly particularly fond of Carr and liked to use him. Carr treated the London office of K2 Intelligence as his fiefdom. Former officials of the firm said he didn't discuss his cases with his colleagues in New York. He also had zero respect or patience, they said, for Jeremy Kroll. When he came to New York, Carr would walk into Jeremy Kroll's office and plant his feet on his desk. He apparently liked to make it clear to the younger Kroll that K2 Intelligence couldn't survive without the revenue he was generating in London and liked to trash-talk Jeremy Kroll behind his back.

BEFORE ROB MOORE MET in 2012 with Matteo Bigazzi he had done one small job for K2 Intelligence. Now Bigazzi would offer him a big job, one that would last for the next four years.

He explained to Moore that a new client had hired K2 Intelligence to find out if a coalition of activists campaigning to ban asbestos use worldwide was being secretly funded by lawyers in the United States. By 2012, the United States, Britain, and other Western countries had banned all types of asbestos. But in Southeast Asia and other developing regions, the use of one variety of the material known as chrysotile or "white" asbestos was continuing and, in some cases, growing.

Moore said that Bigazzi told him that the public health advocates were exaggerating the dangers of chrysotile asbestos, adding that it was safe to use if it was handled properly. The firm's client was particularly suspicious, Bigazzi added, of a tie between the activists and the plaintiff's lawyers because the woman leading the coalition was the sister of an American attorney who specialized in suing asbestos companies. Typically, investigative firms don't disclose the names of clients to operatives and Moore said that Bigazzi simply described its new client as a "U.S. investor" with holdings in the asbestos industry.

Moore and Bigazzi soon hit on an investigative approach that seemed a perfect fit for the former television producer. K2 Intelligence would hire him to create another kind of prank show. He would approach the activists pretending to be a crusading filmmaker working on a documentary about the dangers of asbestos and, using that disguise, gain their confidence and infiltrate the group. Moore then would regularly report back to K2 Intelligence about the activists' lobbying plans.

TO SET THE OPERATION into motion, Moore turned to a technique long used by comedians and spies—misdirection. He knew that calling the activists to say he was making a film about asbestos might set off alarm bells because none of them would have heard of him. He was also worried about spooking the investigation's target, the lawyer's sister, Laurie Kazan-Allen. So instead of calling her directly, he decided to contact her fellow campaigners and get them to make an introduction.

Moore introduced himself to the activists as a longtime investigative filmmaker who was developing a new television series about all kinds of dangerous industries and products—cigarettes, toxic chemicals, hazardous wastes, etc.—only throwing in asbestos at the end as an afterthought. "The advantage of going into this world with a bigger agenda than asbestos is that it might make my entry seem less deliberate," Moore wrote Bigazzi in one note. As for Kazan-Allen, he told Bigazzi, he wanted to interact with her "in the most genuine and heartfelt way possible so that I can establish both an intellectual and emotional connection."

The strategy worked. Health experts associated with the coalition not only encouraged Moore to focus his attention on asbestos, they put him in touch with Kazan-Allen. The activist, who was in her mid-sixties, was so taken by Moore's portrayal of a truth-seeking filmmaker she soon began to see him as a future leader of the ban-asbestos movement. "He was very polite, incredibly engaging, and completely believable," she would say years later.

Moore first got her to invite him to a meeting of anti-asbestos activists in Brussels. Kazan-Allen assumed he wanted to go there to make connections for his film. Instead, Moore went to find out whether her brother and other American lawyers were funding the event. Before long, he traveled to Thailand to attend an occupational health conference after Bigazzi told him that K2's client wanted him to monitor the event because officials there were considering an asbestos ban.

TO AVOID CREATING AN electronic paper trail, Rob Moore and Matteo Bigazzi communicated using an email technique known as "foldering." To use the method, two or more people create a Gmail account to which they know the password. Then, instead of sending emails to each other, they write emails and store them in the account's draft folder. Anyone with access to the account can read the draft email, delete it, and respond by writing another email and storing it as a draft. Over time, Moore and Bigazzi exchanged dozens of draft emails that way, many in which Moore described the progress of his spying. The Gmail account Moore created was called "benthiczonesolutions.com." It was a reference to an ecologic region known as the benthic zone, that exists at the deepest level of a lake, an ocean, or another body of water. It is a place that is pitch-black, oxygen depleted, and inhabited by bottom feeders.

ROB MOORE WOULD SAY he began questioning his assignment even before going to Thailand. He had found nothing to suggest that

American lawyers were funding the coalition and he had concluded based on information the activists had shared with him that chrysotile asbestos was deadly and should be banned. "I don't see the work of an arch mastermind who is unreasonably using disingenuous statistics," he wrote Matteo Bigazzi. "I see the work of campaigners who have a good argument on their side."

At this point, most people would have parted ways with K2 Intelligence. Moore didn't. It may have been the money he was making or the sense of power that comes from spying. He insisted that he remained in the firm's employ for another reason—his Buddhist faith.

Moore said he met with fellow Buddhists to discuss whether he should stay at K2 Intelligence or quit. They told him that if he quit, the firm would simply find someone else to replace him, but that if he stayed, he could potentially do something worthwhile by exposing the asbestos trade. "I could absolutely do this from a Buddhist perspective so long as I didn't do any harm," Moore said.

Whatever his reasons, Moore kept collecting checks from K2 and misleading the activists. Soon he was also duping officials at the World Health Organization after Matteo Bigazzi told him that the firm's client wanted to know if the agency planned new actions to promote an asbestos ban. Moore met with WHO officials and convinced them to help underwrite a short documentary he wanted to make about asbestos use in India. He traveled to Mumbai with a director friend from his time in television and they made a nine-minute-long film titled "Victims of Chrysotile Asbestos." It was hardly an advertisement for the asbestos industry. The film featured portraits of people who were sick or dying from asbestos exposure. "I am saddened that asbestos continues to be used in Southeast Asia," one expert says in the film. "When clearly alternative products are available."

When Moore showed the film to the activists, they loved it.

Bigazzi had a different reaction. Halfway through the film he asked Moore to turn it off. "He said he had seen enough," Moore said.

Just what Bigazzi was telling K2 Intelligence's client about Moore was anyone's guess. But one day when Moore was leaving the firm's offices, a K2 Intelligence operative pulled him aside. He told Moore that the firm worked for questionable people. And there was something else—its client in the asbestos case wasn't a "U.S. investor."

CHAPTER 3

"OPPO"

When Glenn Simpson and Christopher Steele first met in 2010, they couldn't have foreseen how their futures would intertwine. But even back then, there were similarities in the trajectories that had taken them into the investigative business.

Both had left their chosen professions at the same age, forty-five, much earlier than either of them had anticipated and short of their ambitions. As a reporter, Simpson loved the cloak-and-dagger world that Steele had inhabited and was obsessed with the former spy's area of expertise, Russia. Steele had once thought about becoming a journalist and had stumbled into his life as a secret agent while pursuing that possibility. At Cambridge University, he had worked on the school's newspaper and responded after graduating to an ad for people interested in journalism jobs abroad. When he arrived for his interview, he was greeted by MI6 officers seeking new recruits, according to an account in *The New Yorker*.

Steele was fluent in Russian and, after training, he was sent in 1990 to Moscow and worked undercover as a midlevel diplomat at the British embassy, a commonplace cover job for spies.

He hoped to remain in the field but MI6 was rocked in 1996 by a scandal. When Steele was stationed in Paris, a list containing the names of more than one hundred MI6 operatives, including Steele's, was posted on the internet. He was pulled back to MI6 headquarters, a citadel-like building in London with triple-glazed windows that looks like it was built from bomb-proof Lego blocks.

In time, he was given a job on the agency's Russia desk where he focused on the activities of politicians and oligarchs in Russia and Eastern Europe and their associations with organized crime groups. In 2007, when Simpson was writing about Russian mobsters for *The Wall Street Journal*, Steele was serving as a liaison between MI6 and U.S. officials on Russian intelligence matters and spoke that year at an international law enforcement officials conference. Steele had hopes of rising further in the spy agency hierarchy but apparently clashed with its head.

GLENN SIMPSON AND CHRISTOPHER Steele were introduced by a mutual acquaintance, Alex Yearsley, who had also just become a private operative. Previously, Yearsley had worked as the director for special projects at Global Witness, an anticorruption group based in London whose financial supporters included the billionaire investor George Soros. Global Witness has long specialized in exposing bribes paid to politicians in developing regions such as Africa by companies eager to exploit a country's oil, mineral wealth, or other natural resources.

Simpson had gotten to know Global Witness during his time as a *Journal* reporter in Europe. The organization frequently collaborated with news organizations and, in the mid-2000s, it began investigating an obscure company in Ukraine that was involved in the operation of a major natural gas pipeline that supplied energy to Western Europe. Global Witness suspected that a top Russian gangster named Semion Mogilevich secretly controlled the firm and the group was sending leads to prosecutors at the U.S. Justice Department. In 2006, Simpson wrote a front-page article for the *Journal* about Mogilevich and the pipeline firm. "The U.S. is worried that the Russian mafia will spread its influence in the energy industry and use its natural-gas profits to increase its economic and political clout," he reported.

Glenn Simpson and Sue Schmidt had real advantages over other private operatives. Along with their journalistic reputations, they knew from experience about the types of stories that got a reporter's pulse racing and the kind of information needed to put an article together. Still, like any new business, they faced challenges in getting their young firm, SNS Global, off the ground. To scrimp on expenses, they worked out of their homes and the company had only one full-time employee, Margo Williams, an expert in database research who had worked for *The New York Times* and *The Washington Post*.

Simpson hoped to keep a hand in journalism. He and Schmidt went to the New York headquarters of ABC News to see the network's investigative correspondent, Brian Ross, and his producer, Rhonda Schwartz. Simpson proposed an arrangement under which the network would pay SNS Global on a freelance basis for stories based on cases it had handled for clients. Ross and Schwartz got plenty of tips from private operatives but they never paid for them. Simpson's idea was fraught with potential conflicts so they asked an ABC executive who oversaw ethical standards to join the meeting. The proposal never got off the ground and Schwartz, who had known Simpson for years, said he seemed to hate the idea that journalists might now view him as a hired gun. "He wanted people to still see him as a reporter," she said.

LAW FIRMS ARE A major source of work for investigative firms and, to land assignments, Simpson and Schmidt made pitches to attorneys. Almost immediately after leaving *The Wall Street Journal*, they got a job from a lawyer in the Washington, D.C., office of Baker-Hostetler, Mark Cymrot, who had represented Rakhat Aliyev, the ex-Kazakh official about whom Simpson and Schmidt had written articles. Cymrot enlisted SNS Global to work on a lawsuit where he was representing a newspaper publisher in Puerto Rico. The publisher claimed that the island's governor, as a way of retaliating for newspaper editorials critical of him, had cut off public contracts to another company the publisher owned, a claim the politician denied.

There are advantages when a lawyer hires a private operative to work on behalf of a client, rather than having a client hire an investigator directly. Among other benefits, investigative reports that go through a lawyer are shielded from discovery during litigation under the so-called attorney "work product privilege". For a hired spy, there is also a cosmetic benefit to having a law firm as an intermediary. Christopher Steele, during his days at MI6, spied on Kremlin-connected oligarchs. As a private operative, he would work on behalf of Oleg Deripaska, the type of Kremlin-connected oligarch on whom he once spied. The ex-MI6 agent later insisted he met Deripaska only once and that his client was the oligarch's London lawyer. Technically, that was true. Realistically, it was absurd to think that Steele didn't know he was getting paid to advance Deripaska's interests.

ALONG WITH WORKING ON lawsuits, Sue Schmidt became involved in the type of investigation she might have worked on as a reporter, one that looked at how the Chinese government was luring Chinese-American students with advanced degrees back home, to take advantage of the knowledge they had acquired in the United

States. The database expert, Margo Williams, spent much of her time working with "Inside Google," a blog that billed itself as a watchdog monitoring the giant information company's tactics.

For his part, Simpson was quickly drawn back into his comfort zone, one filled with international players engaged in intrigue. Soon he started working on a case with plot lines similar to the Kazakh feud, except that this battle was occurring in a place even fewer people had heard about—Ras al Khaimah, the northernmost of the seven emirates that comprise the United Arab Emirates. RAK was best known for three things: its oil wealth, its proximity to Iran, and its high-quality toilets and bathroom fixtures, sold under the brand name RAK Ceramics.

The clash in RAK involved two sheiks. The older one was in line to succeed his dying father as RAK's ruler. The trouble was that his father preferred his half brother. The spurned sheik fled RAK and hired American lawyers, lobbyists, and consultants who conducted the same type of campaign that operatives had employed on behalf of Rakhat Aliyev. Paid advocates lauded him as the "pro-Western" force in RAK and his face appeared in advertisements on the sides of Washington, D.C., buses that read, "Thank you, America, our people will soon be safe, secure and prosperous again."

Simpson was paid $40,000 by a lobbying firm that represented the deposed sheik to write a report that was circulated to lawmakers and reporters. In it, Simpson warned about the dangers posed by RAK's rulers, claiming that they had ties to arms dealers and terrorists and were helping Iran evade sanctions by allowing Tehran to use RAK as a transshipment point, claims the emirate denied.

While advocating on behalf of RAK's would-be ruler, Simpson took a step he would not repeat for any future client—he registered as a lobbyist. When a reporter asked him in 2009 why he had done so, he responded he had acted "out of an abundance of caution" after consulting with a lawyer who specialized in lobbying rules. "Research firms are obligated to register if their work is used in meetings with government officials," he said.

ON ONE SUNNY DAY, Glenn Simpson and Sue Schmidt were enjoying lunch at a rooftop restaurant of a Washington, D.C., hotel with another potential client, the investor Bill Browder. Since Simpson's initial meeting with him in Brussels, the businessman's life had been turned upside down and he was in the United States on a lobbying mission of his own.

As Browder recounted in his best-selling memoir, *Red Notice*, corrupt Russian officials broke into the Moscow offices of Hermitage Capital soon after his banishment. There, they stole documents which were used to engineer a fraud in which $230 million in Russian taxes paid by Browder's firm were rebated back to them. The scheme made it look like Hermitage Capital hadn't paid taxes and a Moscow court later convicted Browder in absentia of tax evasion.

But years before that happened, a tragedy occurred. A tax lawyer working for Hermitage Capital, Sergei Magnitsky, who had been trying to unravel the tax scam, was arrested by Russian authorities. After spending nearly a year in a Moscow jail, he died in 2008 under troubling circumstances. Magnitsky, who was suffering from an inflamed pancreas, didn't get proper medical treatment while imprisoned and Browder was convinced that the accountant's Russian jailers had beaten him prior to his death.

The investor wanted retribution and had come to Washington in 2010 to lobby Congress to pass a law that would impose financial sanctions and other penalties on Russians involved with Magnitsky's treatment. He also wanted to know who had financially benefited from the Hermitage Capital theft.

During their luncheon, he, Simpson, and Schmidt discussed the possibility of hiring SNS Global to help him trace the stolen money. It seemed like a perfect assignment for Simpson's talents, one that would involve a forensic dissection of shell companies, offshore banks, and shady investment vehicles. In the end, Browder decided to have his own staff at Hermitage Capital conduct the search and the pleasant lunch that Simpson and Browder shared that sunny day was likely the last time they shared friendly words.

BEFORE LONG, GLENN SIMPSON and Sue Schmidt faced a far bigger problem than landing new clients. A little more than a year after its founding, SNS Global was imploding. The two ex-reporters had different personalities and those differences became more evident the longer they worked together. Schmidt was more straitlaced than Simpson and her politics veered toward the conservative. He disliked politicians of all stripes and was far more ambitious than her.

Their business model of conducting investigations only for worthy clients was doomed from the start. It is the controversial players who are in constant need of the services of private operatives and who pay top dollar. In later years, Simpson would work, sometimes for free, for gay rights groups, anti-gun organizations, and other advocacy causes. But he understood where the money was and, unlike Schmidt, he wanted to get rich from being an operative-for-hire.

By mid-2010, the little goodwill left between them had evaporated and they dissolved SNS Global. Afterward, Margo Williams went back into the news business and, in time, Sue Schmidt would turn to consulting and, eventually, work as a freelance journalist. But years later, their experience of working with Simpson was apparently so unpleasant that neither one of them wanted to talk about him again.

GLENN SIMPSON SOON STARTED his new firm, Fusion GPS, with Peter Fritsch, another former *Wall Street Journal* reporter whose ethical compass was more aligned with his own. Fritsch had spent twenty years working for Dow Jones, first as a reporter for its financial wire service, before jumping to the *Journal* as a reporter and an editor. There, he worked in, among other places, Mexico City, Southeast Asia, and Brussels.

Fritsch was a year older than Simpson, and had brown hair and a stocky build. He bore a slight resemblance to actor Nick Nolte and had grown up in Massachusetts, where his father was a minister. He had allies at the *Journal* but, unlike Simpson, there were people who were repelled by him. Fritsch carried himself with the macho swagger of a risk-taking foreign correspondent and came across to those who disliked him as a know-it-all who could quickly turn bullying or condescending. He also had a hair-trigger temper and was prone to making inflammatory comments.

Early in his career, a female editor wanted to get rid of him because she found him so abrasive but male colleagues arranged for him to receive a transfer. Later, when Fritsch was a senior editor, a reporter charged that he had used a racial epithet when describing to the reporter how he needed to act subserviently to his superiors if he wanted to get ahead at the paper.

In 2004, that reporter, Shawn Crispin, whose job was then being eliminated, notified lawyers at Dow Jones, the *Journal*'s parent company, that he had a tape recording of Fritsch's remarks.

"Going forward, I suggest that you ask Mr. Fritsch whether or not my representation of our [conversation] is accurate," he wrote in an email.

More than a decade later, a spokesman for the *Journal* declined to comment on the episode. Fritsch also didn't respond to questions about it. But a former Dow Jones newsroom manager who was in Asia at the same time as Fritsch, backed up Shawn Crispin's account.

AT FUSION GPS, GLENN Simpson and Peter Fritsch played complementary roles. Simpson was scattered and disorganized. Fritsch had management skills and oversaw the business as it grew. The firm's offices were located in a triangular-shaped building above a Starbucks on Connecticut Avenue, about halfway between Dupont Circle and the Washington Hilton, the hotel where the assassination attempt on President Ronald Reagan had occurred. In time, it grew to employ about twelve people, making it a small company by investigative industry standards. Unlike competitors, Fusion GPS didn't strive to affect a corporate tone. Instead, its offices resembled a frat house in need of deep cleaning. Employees sat balanced on blow-up balls in front of computers and an open bar was stocked with liquor bottles. The Jolly Roger flag was pinned to the wall of the office's kitchen and, later on, framed tweets by Donald Trump attacking Fusion GPS lined a wall.

Fritsch filled another niche at Fusion GPS. He played bad cop to Simpson's nice guy. It was a job he seemed to relish, particularly when his former colleagues at *The Wall Street Journal* started investigating the firm's clients.

IN 2012, GLENN SIMPSON got involved in another lucrative arena for private operatives, one that reached a new high-water mark with the dossier. It was called political opposition research, or "oppo."

Political candidates have always searched for episodes in an opponent's past—marital infidelities, arrests, etc.—to use against them. For decades, that job was handled by campaign staffers and volunteers who hunted for information in newspaper archives and court records. Then private spies entered the game and an amateurish pursuit was turned into a weapons-grade enterprise.

If there was a moment when hired spies jumped into politics to stay it came in 1992, when Bill Clinton ran for president. An investigative reporter for *The Washington Post*, Michael Isikoff, got a tip that the Democrats had hired private operatives to squelch rumors that Clinton had a history of extramarital affairs. One of his reputed paramours had already given an interview to a tabloid newspaper and other women were set to make similar allegations about him.

Isikoff learned that the Democratic National Committee had hired a San Francisco–based investigative firm to dig up dirt to discredit the women or to get them to sign affidavits attesting they never had an affair with Clinton. Isikoff looked for payments to the firm on filings that the Clinton campaign was required to make with the Federal Election Commission. But he didn't find any and the journalist soon discovered the reason: Clinton campaign officials had come up with a technique to conceal them. They had disguised the investigators' fees by including them in funds paid to a law firm that then passed the money on to the operatives. A spokeswoman for Bill Clinton's campaign defended its use of private investigators, saying it protected the candidate against what she described as "bimbo eruptions." As it turned out, the San Francisco–based private investigative firm, Palladino & Sutherland, dispatched by the Clinton campaign to deal with those "bimbo eruptions" was hired three decades later to dig up dirt on women accusing Harvey Weinstein. By then it had rebranded itself to reflect changing times. It was now called PSOPS.

WHILE HE WAS A journalist, Glenn Simpson had railed against the presence of spies-for-hire in politics. In 1996, just before joining *The Wall Street Journal*, he coauthored a book titled *Dirty Little Secrets* with an expert on elections, Larry Sabato, a professor at the University of Virginia. They described *Dirty Little Secrets* as an investigation into the growing corruption of American electoral politics. Campaigns are "sinking ever deeper into a bog of sleaze and slime—a primordial political ooze whose toxicity is increased by new technologies that make voters who are already turned off hate politics all the more intensely," they wrote.

Simpson gave special attention in the book to one hired operative who had helped pioneer the involvement of private investigators in oppo. He was Terry Lenzner, a well-known lawyer in Washington who had once drawn the admiration of journalists and public interest advocates. Lenzner had initially worked at the Justice Department as an attorney on high-profile civil-rights cases. But he became best known for his role on the Senate Watergate Committee and the hearings that led to the resignation of President Richard Nixon.

During the height of the 1980s corporate takeover boom, Lenzner founded a corporate intelligence firm called the Investigative Group International and capitalized on his Watergate fame. IGI, as the company was known, quickly grew from three employees to one hundred and successfully competed against Kroll Associates for corporate work. Lenzner had started doing oppo work on behalf of Bill Clinton back when Clinton was the governor of Arkansas and, a decade later, when the Monica Lewinsky affair exploded, he was still at it and looked for dirt about the White House intern.

Personality-wise, Lenzner and Glenn Simpson appeared to have a lot in common. They cherished their outsider status but they were really Washington insiders drawn to power. And they shared another trait as well: if they did something they would have considered wrong when someone else did it, it wasn't wrong when they did.

GLENN SIMPSON GOT HIS start in his oppo career in 2012, when President Barack Obama's reelection campaign hired Fusion GPS to muddy up Mitt Romney, his Republican opponent. Simpson met with journalists at *The New York Times* and other publications to shop stories about Romney's time as head of a large private equity firm, Bain Capital. The firm had profited by laying off scores of workers at companies and Romney, who was a multimillionaire, had used offshore tax havens to shelter his wealth. The Obama campaign would make Romney's wealth and time at Bain Capital lynchpins of its attacks against him but Fusion GPS soon tipped its hand in that attack because of an amateurish mistake.

The episode unfolded when a clerk at a courthouse in Idaho Falls, Idaho, received a phone call from someone in the summer of 2012 seeking copies of lawsuits filed there against a local businessman who was a major Romney donor. Several weeks earlier, an Obama campaign website had posted a press release attacking the businessman, Frank VanderSloot, and several other prominent Romney supporters.

In the press release, VanderSloot, whose company sold dietary supplements, was described as "litigious, combative and a bitter foe of the gay rights movement." The release linked to an article in *Mother Jones* magazine about VanderSloot, which reported that he tried to pressure an Idaho public television station into dropping a children's program that discussed homosexuality. The article also said that the businessman had gone after a local reporter, who was gay, over articles about the failure of the Boy Scouts to investigate troop leaders suspected of being pedophiles.

VanderSloot, who denied the charges, didn't know at first who was looking for lawsuits against him. A person named Michael Wolf had called the court and when the businessman did an internet search for him he was listed as an intern on a Senate committee. That discovery alarmed VanderSloot because his company had faced previous scrutiny and he feared that congressional Democrats might be investigating it again because of his Romney support.

He contacted a conservative columnist for *The Wall Street Journal*, Kimberly Strassel, and alerted her about what had happened. Strassel started making phone calls and quickly discovered that Wolf no longer worked on Capitol Hill but at Fusion GPS. He was still learning the ropes because when he sent follow-up faxes to the Idaho courthouse, he didn't send them from a generic fax machine at Staples but from one at the investigative firm.

Strassel wrote a column about the episode, pointing out that Fusion GPS's website described it as providing "strategic intelligence" and expertise in politics. "That's a polite way of saying 'opposition research,'" she wrote. She called Simpson to interview him but he didn't call back, sending an email instead. It read, "Frank VanderSloot is a figure of interest in the debate over civil rights for gay Americans." But Simpson was apparently so embarrassed by the incident that he told a friend he took down Fusion GPS' website for a time.

AFTER THE 2012 ELECTION, VanderSloot filed a lawsuit against *Mother Jones* claiming that the magazine had defamed him. And to gather evidence for it, his lawyers subpoenaed Simpson, seeking to interview him under oath about his interactions with *Mother Jones* and Obama officials. Simpson insisted in filings that he wasn't involved with the *Mother Jones* piece but skirted the question of ties between Fusion GPS and the Obama campaign.

His lawyers also turned to a tactic that attorneys for Fusion GPS would employ again years later when the firm was sued over the Trump dossier. They sought the shelter of a statute designed to protect public interest groups and activists from retaliatory lawsuits filed by companies or wealthy individuals who use costly litigation as a weapon to silence critics.

Dozens of states have passed laws to protect citizens against such actions, known as "SLAPP" lawsuits, shorthand for "strategic lawsuits against public participation." Spies-for-hire were probably not on the minds of lawmakers when anti-SLAPP statutes were passed but Simpson's lawyers maintained that he qualified for protection. The case's judge didn't buy the argument but Simpson was saved from testifying when VanderSloot's lawsuit against *Mother Jones* was dismissed.

YEARS LATER, AFTER THE Trump dossier became public, it would be revealed that the payments received by Fusion GPS to work on behalf of Hillary Clinton's campaign didn't appear on federal filings for the same reason that payments to investigators working for Bill Clinton were invisible – the funds were paid out by a law firm. In his 1996 book, *Dirty Little Secrets*, Simpson had recounted Michael Isikoff's article in *The Washington Post* about how the Clinton campaign had hidden its payments to private operatives so they wouldn't appear on public filings. "In a sense, the effort to hide snooping activities is a good sign," Simpson wrote. "Apparently, some people still have a sense of shame about engaging in such conduct, or at least a fear of what the public will think."

Simpson and Isikoff were close friends who partied together and cheered on each other's professional exploits. "Mike will pull your fingernails out over coffee discussing lawn care," Simpson told *The New York Times* in 2005. "He is just a born interrogator."

But Isikoff was taken aback when he learned that Simpson had agreed to use a middleman which masked payments to Fusion GPS. When he asked his friend why he had done so, Simpson's response was simple. "It's legal," he said.

CHAPTER 4

THE LONDON INFORMATION EXCHANGE

LONDON, 2014

In the world of private spying, information is currency and operatives scoop up whatever material they can find from wherever they can find it and then turn that information into cash by selling it to clients or using it as bait to attract new ones. Many operatives aren't too particular about the provenance of the information in which they trade. In some cases, it is legitimately obtained. In other cases, it may have been stolen or hacked.

Lawyers who get caught trafficking in purloined documents can get disbarred but few attorneys are stupid enough to let that happen. Instead, lawyers take information obtained by hired operatives and use it as "intelligence" to determine, amongst other things, which type of records to legally request from an opponent or how best to attack a witness. Hedge funds are also big fans of hired spies and use information they gather to bet against companies in the stock market.

London, with its legions of investigative firms, serves as a center of the informational trade as well as home to its intermediaries and middlemen. And one particularly active player over the past two decades in that marketplace was Mark Hollingsworth. Like Rob Moore, Hollingsworth worked as a contractor for investigative firms though his roster of clients was far more extensive. Hollingsworth also had a special skill. While Moore could pretend to be a journalist, Hollingsworth actually was one and he operated for years as a hybrid version of a reporter and a private operative.

THE SOFT-SPOKEN HOLLINGSWORTH, WHO was born in 1959, had thinning hair and stood with a slight hunch that tall people unconsciously adopt to make themselves appear shorter. He also had a bland, downbeat manner reminiscent of Eeyore, the depressed donkey in the Winnie-the-Pooh stories. Hollingsworth was always eager to share information with other journalists and younger reporters looked up to him. He was a pack rat and compiled reams of files over the years about various oligarchs, including a database that listed the owners of properties throughout London. He was also a frequent visitor to the offices of Global Witness, the anticorruption group, where he traded tips with investigators.

Hollingsworth's dual career as a journalist and private operative would have been short-lived in the United States. Apart from *The National Enquirer* and other tabloids, American newspapers don't buy information from private investigators. And if a major outlet discovers that a staffer or freelancer is moonlighting as a hired spy, he or she will be fired and blacklisted.

The culture and practices of the news media in England have long been different. In his day, Ian Withers, the private eye, said that he worked decades ago as a contract investigator for a major British newspaper, the *Sunday Times*, and would feed information he had gathered to investigative reporters at the paper, where it appeared under their bylines. He claimed he equipped one *Sunday Times* journalist with a special Samsonite briefcase that an engineer at his firm had legally tricked out to conceal a tape recorder. When the briefcase's handle was held upright, the tape recorder remained off. When the reporter put the briefcase down on the floor, the handle fell to the side and a recording started.

The widespread use of private eyes by British newspapers came to light in 2011 when it was revealed that operatives working for a tabloid owned by Rupert Murdoch, *The News of the World*, had hacked into the cellphones of a murdered schoolgirl and the victims of a London terrorist attack. The scandal led to public hearings in England but afterward practices at most British newspapers didn't change completely. Some publications adopted policies that asked freelancers to disclose potential conflicts of interest. But it was a voluntary process with scant oversight and one that created ample opportunities for Hollingsworth and other reporters/operatives with feet in both worlds. In England, there is even a name for them—they are called "tame" journalists.

MARK HOLLINGSWORTH FREELANCED FOR *The Guardian*, the *Financial Times*, and other British newspapers, writing articles about oligarchs, corporate chicanery, and corrupt politicians. He consulted with the British Broadcasting Corporation on investigative pieces and authored several books, including one titled *Londongrad*, which chronicled that city's invasion by wealthy businessmen from Russia and other parts of the former Soviet Union.

Then, his life as a journalist and a private operative began to join with the fortunes of a company that had mines in Kazakhstan and Africa. And in following years, his interactions with that company would offer a guided tour of every compromised corner of the corporate investigations industry.

That firm, Eurasian Natural Resources Corporation, or ENRC, had been founded by three oligarchs whom British newspapers referred to as the "Kazakh Trio" or the "Trio." The three men didn't hail from Kazakhstan but they were allies of that country's president and the Kazakh government was a major investor in ENRC, whose shares were once traded on the London Stock Exchange.

In 2011, a scandal involving ENRC erupted when a whistleblower alleged that the company had paid bribes to win mining rights. Facing scrutiny, the firm, which has denied any wrongdoing, hired a London-based lawyer to conduct an independent inquiry into the allegations but that didn't go so well. The attorney, Neil Gerrard, wrote that he had been given forged ENRC documents to review, had been taken to inspect a fake company office, and couldn't account for millions in company expenditures. And after someone leaked Gerrard's report to a newspaper, his findings came to the attention of the Serious Fraud Office, a British government agency that investigates corruption.

Despite the fact that Gerrard and his law firm denied having done this, the mining company and its owners faced not only the potential for public censure if the bribery allegation was true but also the possibility of paying tens of millions of dollars in fines. Given those stakes, the ENRC case would soon become a battle-field for warring armies of corporate spies, some hired to work for ENRC and the "Trio" and others hired by their enemies.

THE ENRC SCANDAL WAS big news in England and Mark Hollings-worth did more than just write about it. He soon came into posses-sion of a trove of internal ENRC documents apparently courtesy of a computer security expert whom he referred to as "Magic". That expert, Robert Trevelyan, who specialized in cyber-security, counted corporations and private intelligence firms among his cli-ents. "Rob is a highly experienced investigator with a deep knowl-edge of IT disciplines and knowledge," his corporate bio stated. ENRC retained Trevelyan over the years for a variety of computer-

related jobs and he had assisted the lawyer hired to review the whistleblower allegations. During that project, Trevelyan made copies
of ENRC hard drives that contained thousands of internal emails,
reports, and other documents. The trouble was, ENRC would claim
years later, he walked away with those documents when the project
was over and joined forces with Mark Hollingsworth.

BEFORE LONG, HOLLINGSWORTH WAS allegedly shopping internal
ENRC records to its competitors, enemies, and other hired spies.
One of his supposed partners in that enterprise, ENRC would
later claim, was Alex Yearsley, the former Global Witness investi-
gator turned private operative who introduced Glenn Simpson to
Christopher Steele. Hollingsworth and Yearsley were old friends
and, in 2011, they allegedly approached one possible customer for
ENRC-related documents, the company contended. It was a rival mining company engaged in a lawsuit seeking $2 billion from
ENRC. The lawyer declined to accept the ENRC documents.
Hollingsworth and Yearsley supposedly next went to a corporate
intelligence firm hired by the company suing ENRC. It wasn't interested, either, but Hollingsworth had better luck with another
old friend—Glenn Simpson.

HOLLINGSWORTH AND SIMPSON MET in the mid-2000s through a
shared interest in Russian organized crime and, in 2009, the British
journalist/operative began working as a contractor for SNS Global,
Simpson's first firm. By 2011, his work for Simpson and Fusion GPS
was ramping up, thanks in part to his stash of ENRC documents.

At that time, emails show, Simpson was pitching the services
of Fusion GPS to a prospective client interested in ENRC. That
client's identity wasn't disclosed in the emails but it was likely a
hedge fund or another investor because ENRC shares were then
publicly traded.

In mid-2011 Hollingsworth met with Simpson in London to discuss the ENRC documents. Afterward, he sent Simpson a follow-up email outlining some one hundred categories of information related to ENRC, its owners, and its executives that were available on the "disc."

"In my view, this is pretty devestating [*sic*] and hopefully you agree!!" Hollingsworth wrote.

Fusion GPS, it appears, landed the account because Simpson later contacted Hollingsworth: "I have to brief client orally on Monday in DC," the Fusion GPS operative told him. "Need a current roundup."

Hollingsworth told Simpson they were in luck. "Our mutual friend Magic has obtained new documents," he wrote.

A FEW YEARS LATER, another player in the investigative industry's underground who knew Glenn Simpson and Mark Hollingsworth arrived at London's Heathrow Airport. He was carrying a computer drive loaded with documents passed to him about a company called International Mineral Resources, which was also controlled by ENRC's founders, the Kazakh Trio.

That man, Rinat Akhmetshin, had lived since 1994 in Washington, D.C., where he worked as an operative-for-hire and a political jack-of-all-trades. At times, Akhmetshin, who was born in Russia, lobbied for groups with ties to his homeland or countries in Eastern Europe. At other times, he brought cases to American law firms from clients in Russia or the former Soviet republics. At still other times, he engaged in bare-knuckled smear campaigns or served as a middleman who supplied journalists with documents of questionable provenance.

Akhmetshin said he charged $450 an hour for his services. He had a simple way of describing the unifying principle connecting his various pursuits. He got paid, as he put it, "to fuck with people" on the other side of legal or political disputes from the people paying him. Short and squarely built, he rode around Washington on an orange bicycle and cultivated lots of journalists. Reporters liked him, though they rarely relied on anything he said.

Akhmetshin, who was born in 1968, had heavy black eyeglasses and an overgrown crew cut that made him look like the lead character in the 1970s cult movie *Eraserhead*. He held a doctorate in organic chemistry but had long ago left the lab for more lucrative activities. He was amusing, well read, cultured, and reveled in his knowledge of fine wine and food. Lunching with a journalist at a high-end restaurant near the White House, he asked a sommelier to serve him pink champagne in a "burgundy glass" before moving on to a glass of Riesling as a pairing with an appetizer of rabbit pâté.

Akhmetshin was conscripted into the Russian army in 1986 and served briefly in Afghanistan, assigned to a unit involved with military counterintelligence. He insisted he had never spied for the Kremlin but some private operatives believed that he was only able to operate in Russia and Eastern Europe as freely as he did with the blessings of real spies who could then call on him when they wanted a favor returned. He also maintained he never was involved in computer hacking because he didn't know enough about computers to know how to hack them. That said, it wasn't unusual for him to turn up to a meeting with a journalist carrying a thumb drive containing emails or computer files.

IN 2017, RINAT AKHMETSHIN would earn an unwanted turn in the public spotlight when it was revealed that he had taken part in the infamous 2016 meeting at the Trump Tower where Donald Trump Jr. expected to receive dirt from Russians about Hillary Clinton.

But his 2014 trip to London had a more prosaic purpose—he was trying to cash in on documents taken from International Mineral Resources, the company connected to the Kazakh "Trio."

Akhmetshin's entanglement with International Mineral Resources, or IMR, had started two years earlier, in 2012, when American lawyers representing a company suing the Trio-connected firm retained him as a consultant. The company, EuroChem, was Russia's largest fertilizer producer and it had brought an action against IMR in Europe, claiming that one of the company's units had bribed a EuroChem manager to buy shoddy drilling equipment that caused a mine to collapse.

Meanwhile, IMR announced that its computers had been hacked and, before long, Akhmetshin was, among other things, showing internal IMR records to reporters in an effort to generate negative news stories about the Trio. On various trips to London, he met with Mark Hollingsworth and reporters working for *The Guardian*, Reuters, and *Harper's Bazaar*. Akhmetshin also dangled the records in front of a Global Witness investigator whom he invited to have breakfast with him at a posh hotel in Mayfair called Brown's, a favorite haunt of private spies.

Akhmetshin's involvement in the IMR case was quite profitable. By mid-2013, a year after he was hired by EuroChem's American lawyers, he had received more than $145,000 for part-time work. But at that point, he said, he was fired by the law firm that hired him. "They dumped me," he said without elaborating.

The official end of his ties to the case didn't mean, however, that he couldn't make more money from it. Akhmetshin said he received a phone call several months after his dismissal from the lawsuit from an investor and lawyer in Israel who said he was considering acquiring assets owned by the Trio in Kazakhstan. At a subsequent meeting in New York, Akhmetshin told the Israeli businessman, Baruch Halpert, that IMR documents were available for sale. They struck a deal and when Akhmetshin arrived in London in early 2014, he was there to hand-deliver a computer drive to Halpert.

WHEN RINAT AKHMETSHIN LEFT his London hotel room for his rendezvous with Baruch Halpert he saw an easy payday ahead. He never realized he was walking into a trap. Private spying is a treacherous business. If an operative-for-hire is given the opportunity to make money spying on a former client, most jump at the chance. They also get paid to spy on each other. There isn't anything personal involved. Like much else in the business, it's just about money.

Earlier, IMR had hired a private spying firm to investigate Akhmetshin's suspected role in the hack. Operatives with that firm, GlobalSource, had been tracking his movements for weeks with a precision that suggested they had an informant feeding them information or were monitoring his emails. GlobalSource knew the precise date of Akhmetshin's arrival in London for his meeting with Baruch Halpert as well as the name of the hotel where he planned to stay. The firm's operatives were staking out its lobby when Akhmetshin arrived there to check in and recorded the time in their notes as 5:30 p.m.

For Akhmetshin, his moment of truth unfolded inside the coffee shop of the Café Royal Hotel, an elegant five-star hotel on Regent Street near Piccadilly Circus. He had arranged to meet Halpert there but when he arrived he didn't see the Israeli businessman so he took a table. About ten minutes later, Halpert joined him. Akhmetshin didn't know it but a GlobalSource spy had slipped into the coffee shop before Halpert's appearance and seated himself at a nearby table.

The GlobalSource operative testified in an affidavit that he overheard Akhmetshin tell Halpert that he had organized the hack of IMR's computers to benefit his client, EuroChem, and had used Russian hackers to do the job. He stated he then watched as Akhmetshin handed a computer drive to Halpert, telling him that it contained about 50 gigabytes of IMR-related data, including "memos, emails and stuff" and "folders and documents."

When Halpert replied that he didn't know how much data was in a gigabyte, Akhmetshin reportedly told him. "There is a lot of shit. There is a lot of the stuff, so . . . but that's why you are paying money."

IMR LATER BROUGHT A proceeding in a U.S. court that accused Akhmetshin of organizing the hack and distributing stolen documents not only to journalists but to the American law firm that represented EuroChem, charges he and the law firm denied. The company asked an American judge to issue an order that would require Akhmetshin to turn over all emails relevant to the episode so that IMR could use them in the actions that EuroChem had brought against it in Europe.

For the proceeding, officials of GlobalSource submitted affidavits describing the firm's operation. The filings were filled with preening about the firm's supposed prowess. They also left several critical questions unanswered, most notably, the name of the person who had decided to drop a dime on Akhmetshin and the role, if any, of Baruch Halpert in the operation. Nonetheless, the firm's statements pulled back the curtain on how private spies operate.

One GlobalSource executive stated that the firm, after getting its assignment from IMR, reached out to its own network of sources in London's informational underground, including people it believed might have gotten the hacked IMR material or who knew people who had. "We requested that they provide us copies of the hacked material information so that we would have proof that Mr. Akhmctshin was in fact distributing stolen documents," he said in his affidavit.

Next, the GlobalSource executive said, the firm circulated a list of locations in London where someone who had obtained the hacked material could drop it off without questions asked. "Because sources are often nervous about turning over sensitive information like this, we told them that they could drop off the hacked information anonymously," the executive testified.

The sentiment, while touching, failed to address why anyone would give GlobalSource something it was getting paid to get without getting paid themselves. Still, the firm's executive went on to explain that he was staying in December 2013 at a London hotel designated as a drop-off site when a concierge called his room to say that someone had left an envelope for him in the lobby. "I picked up the envelope and opened it in front of the concierge," the operative testified. "The envelope contained a thumb drive."

Computer experts who examined the thumb drive for Global-Source reported finding 28,000 files belonging to IMR on it. They also discovered digital fingerprints in the metadata showing that someone with the initials "RA" had opened some files, a discovery that pointed right at Rinat Akhmetshin. The subsequent setup at the Café Royal Hotel was icing on the cake.

RINAT AKHMETSHIN INSISTED WHEN questioned by an IMR lawyer that he had gotten the company's documents from an old client and friend, a former Kazakh prime minister living in London. He adamantly denied hacking the documents or arranging for the hack. Instead, he insisted that he and the ex-Kazakh official routinely obtained corporate records as part of swaps that went on all the time between reporters, private operatives, and others.

"You know, people often ask me for information and there's a thing called the London information exchange bazaar," Akhmetshin testified. "It's almost, like you know, an exchange bazaar where people—people kind of exchange information."

The IMR lawyer seemed baffled. "The London information bazaar," he asked, "is that formal?"

"It's informal," Akhmetshin replied.

"And do you participate in the London information bazaar?" the lawyer asked.

"From time to time," he said.

Akhmetshin would later say he believed that Baruch Halpert had helped set him up. Whatever the case, the American judge ordered him to produce more emails, pointing to ones he had already disclosed where he told EuroChem's attorneys that "the project is up and running and it is churning out info." Soon after the ruling, the action was concluded.

FOR AKHMETSHIN, THE IMR case proved to be a disaster, both professionally and financially. Instead of fucking someone, he had gotten seriously fucked. In time, however, salvation emerged in the unexpected form of Bill Browder.

Following Browder's nonstop lobbying, Congress in 2012 had passed the Magnitsky Act, a law named after the tax specialist, Sergei Magnitsky, who had died in a Moscow jail. Under the statute, sanctions were imposed on dozens of Russian officials involved in the episode. An outraged Vladimir Putin retaliated by suspending a program through which American parents had been adopting Russian infants.

Staffers at Browder's investment firm, Hermitage Capital, also had made progress solving part of the puzzle that the investor had nearly hired Glenn Simpson to figure out—the whereabouts of the fund's stolen money. They traced a portion of the looted

$280 million to a Russian-owned real estate company called Pre-
vezon Holdings, that had apparently used the funds to buy property in Manhattan.

In 2013, Browder brought his findings to the Justice Department and federal prosecutors in Manhattan filed a civil lawsuit
against Prevezon, to seize the company's Manhattan properties.
Akhmetshin got a call from a Russian contact who asked him to
recommend American lawyers to defend the firm and he referred
it to the Washington, D.C., office of BakerHostetler.

SOON THE TEAM ASSEMBLED to defend Prevezon took on the air
of a garage band reunion. The two BakerHostetler lawyers on the
case, Mark Cymrot and John Moscow, had known Glenn Simpson
for years, first as a journalist and then as a private operative. They
hired him to work on the Prevezon case and Rinat Akhmetshin
later joined the team.

For Akhmetshin, the Prevezon case would mark a new start
after the IMR fiasco, and set him down on a path toward the
Trump Tower meeting. And Simpson's metamorphosis from journalist into a private operative would enter yet another phase, one
where he would seek to undercut the reputation of a person he had
once cultivated as a source: Bill Browder.

CHAPTER 5

BAD BLOOD

NEW YORK, 2016

The relationship between private operatives and journalists rarely comes to light and when it does it is usually because things go sideways. One memorable flameout involved Terry Lenzner, the former Watergate staffer and political oppo expert, whose career arc Glenn Simpson's would come to resemble.

In 1996, Lenzner's firm, IGI, was hired to gather dirt about the cigarette industry whistleblower who was featured on the *60 Minutes* exposé that Lowell Bergman produced. His former employer, Brown & Williamson, wanted to undercut the credibility of the ex-executive, Jeffrey Wigand, and IGI was hired to dig up dirt about him. Months later, the firm produced a five-hundred-page dossier about Wigand that claimed that he had, among other things, assaulted his wife, got caught shoplifting, and lied on his résumé, charges he denied.

With the aim of inflicting maximum damage, the IGI dossier was funneled to reporters at *The Wall Street Journal*. But the article published by the paper was very different from the one anticipated. In checking out IGI's claims about Wigand, *Journal* reporters found that many of them were untrue, rumors, or assumptions. "B&W's tactics aren't unheard of in high-stakes litigation," the newspaper reported in a front-page article. "Whatever they turn out to prove about Mr. Wigand, they provide chilling insight into how much a company can find out about a former employee—and the lengths it will go to discredit a critic."

For Lenzner, the episode was a black eye and he cast blame for it on low-level staffers at the firm. Others viewed the case as proof that Lenzner, once lionized as a truth seeker, had fallen from grace. "Everyone has good and bad, and Terry was on balance good, noble," one of his friends told *Vanity Fair* magazine. "Now he is slimy. I see something Shakespearean in all of this."

HIRED OPERATIVES HAVE A long history of trying to intimidate journalists or ruin them. One of the better-known cases involved a notorious private eye in Hollywood named Anthony Pellicano. A reporter for the *Los Angeles Times* decided to investigate Pellicano, who was eventually imprisoned on firearms charges. One day, she found that someone had cracked her car's windshield and put a sign on it that said "STOP." A dead fish with a rose in its mouth lay on a car seat.

The *New Yorker* writer Jane Mayer became the target of a different kind of campaign. In the early 2000s, she discovered that private spies had been dispatched to try to destroy her reputation as apparent payback for an article she wrote about the Koch brothers, the wealthy and politically conservative industrialists. The operatives produced a dossier claiming to show that Mayer had plagiarized the work of other journalists without giving them credit. The charge, if true, would have destroyed her career. "Dirt, dirt, dirt is what the source later told me they were digging for in my life," she wrote in one of her books, *Dark Money*. "If they could not find it they would create it."

The charges against Mayer proved ill-founded and she sought out the head of the corporate investigations firm she suspected was behind the smear campaign. When Mayer confronted him, he exhibited the type of cowardice that private spies usually adopt when they are caught in the act: he declined to comment.

DURING THE 2010S, THE interactions between hired spies and journalists took on a new and interesting twist. With more reporters leaving shrinking newsrooms to become private operatives, ex-journalists ended up investigating those still on the job, including their former colleagues.

One such episode involved the *Los Angeles Times*. In 2013, it published a front-page article by a reporter, Jason Felch, that examined whether an area college was complying with federal rules requiring schools to report alleged sexual assaults on campuses.

Officials at the California school, Occidental College, previously had acknowledged that they had failed in earlier years to notify federal authorities about all assault allegations lodged by students. But Felch charged in the article that the problem was continuing and that Occidental failed to report twenty-seven allegations of sexual assault made by students during the 2012 school year.

Facing an uproar, college officials hired a crisis management firm run by former reporters at the *Los Angeles Times* to investigate Felch's work. One of the ex-journalists had been particularly close to Felch and had collaborated with him on a series of articles for the *Times* about looted antiquities that became the basis for a 2011 book they cowrote called *Chasing Aphrodite*.

For Felch, things quickly fell apart. At a 2014 meeting with top editors at the paper, Occidental officials presented them with the former reporters' findings. They concluded that the college hadn't been required to report any of the 2012 incidents cited by

Felch in his article because none of them involved allegations of on-campus assault or rape. Instead, they were composed of less serious incidents such as claims of inappropriate text messages, about which colleges didn't have to notify authorities.

Felch's journalistic career might have survived that blunder. But as his work was coming under scrutiny, he told his editor about an action that violated a basic journalistic practice. Soon after he started investigating the school's assault reporting policies, he began a romantic affair with a professor there who became a source for him. Felch later said the relationship began after he wrote his first article about Occidental College. Whatever the case, the *Los Angeles Times* was forced to run a major correction about Felch's article and the paper's managing editor stated in an accompanying note that the reporter was being fired for failing to disclose the relationship. The editor described that omission as "a professional lapse of the kind that no news organization can tolerate."

Fortunately for Felch, he soon landed on his feet. A few months later, he was hired by Fusion GPS.

IT WAS PETER FRITSCH who handled the firm's more combative encounters with reporters. It was a job that suited his temperament. Not long after BuzzFeed posted the Steele dossier, Brian Ross, the ABC News correspondent, reported that an unnamed source had disclosed that Hillary Clinton's campaign had paid Fusion GPS about $1 million for its work. Shortly afterward, Ross got a call from an irate Fritsch, telling him the figure was wrong and demanding to know the name of the person who had provided it.

Ross didn't respond because Glenn Simpson had given him the information with the understanding that Ross could use it so long as he didn't identify it as coming from him. Fritsch became so angry that he next contacted a senior executive at ABC News to demand that the network run a correction, but that didn't happen.

WHEN DEALING WITH REPORTERS, Peter Fritsch liked to wave around his experience at *The Wall Street Journal* as though it were a talisman that would dispel the fact that he was now a hired gun serving the interests of paying clients. And one client for whom Fritsch would go to bat against his former *Journal* colleagues was a little-known utility company in Venezuela called Derwick Associates.

In 2014, when Fusion GPS was hired to work on behalf of Derwick, U.S. authorities were investigating the company for possible bribery and money laundering. Derwick retained an American defense lawyer named Adam Kaufmann, who once worked alongside John Moscow in the Manhattan District Attorney's Office. He hired Fusion GPS and Fritsch handled the Derwick case because he was a fluent Spanish speaker and had spent years for the *Journal* in Mexico City and South America.

Derwick was run by three young Venezuelan businessmen who had no prior experience operating power plants. What they did seem to have were serious connections to the country's socialist government and reportedly received no-bid contracts from it. Federal investigators suspected that Derwick was kicking back money to Venezuelan officials through a complex currency exchange scheme.

Reporters at *The Wall Street Journal* had launched an investigation into Derwick and, in 2014, one journalist on that team, Jóse de Córdoba, arrived in Caracas to interview the firm's chief executive. When he walked into the executive's office, Peter Fritsch was waiting to greet him.

The two men knew each other. The reporter was based at the *Journal*'s bureau in Mexico City when Fritsch had headed it. During that time, they had enjoyed a friendly relationship and socialized together. Fritsch, like other *Journal* staffers, knew that the Cuban-born de Córdoba had come to the United States as a child when his parents fled their homeland after Fidel Castro seized power.

Both Fritsch and Adam Kaufmann insisted to de Córdoba and other *Journal* reporters that Derwick hadn't done anything wrong and was helping U.S. authorities probe corruption in Venezuela. Lawyers and "crisis management" experts make good money every day defending their clients to journalists. But things soon got personal. Fritsch argued to one *Journal* staffer that de Córdoba, given his family's experience in Cuba, might be biased against Derwick because of the firm's ties to Venezuela's socialist regime. Some of de Córdoba's colleagues found Fritsch's tactics distressing and de Córdoba felt Fritsch was trying to mislead him by steering him away from investigating Derwick. "Jose once thought Peter was a fantastic journalist," a *Journal* staffer later said. "He was upset about what he had become."

IN 2015, NOT LONG after the Derwick incident, Peter Fritsch contacted another *Journal* reporter, John Carreyrou. He had been part of a team of reporters at the *Journal* who had recently been awarded a Pulitzer Prize for a series of articles that exposed how doctors were draining billions of dollars from Medicare. It was the paper's first Pulitzer for reporting since Rupert Murdoch had acquired it a decade earlier and a hopeful sign that its tradition of tough journalism was returning.

"First, big congrats on the big P . . . Has Rupert had you on his yacht yet?" Fritsch wrote Carreyrou in May 2015, a month after the Pulitzers were announced. "I'm Glenn's business partner here in D.C. been a lot of fun so far. Had something come my way and wanted to ask."

Carreyrou and Glenn Simpson had been based in Europe at the same time for the *Journal* and occasionally worked together. His previous dealings with Fritsch had been minimal but he said he would be happy to help out if he could.

Fritsch explained that Fusion GPS was working on a project related to the medical laboratory testing industry. He said in a note that he just had spoken with an industry whistleblower who had been involved in lawsuits that charged the testing industry's two corporate giants, Quest Diagnostics and LabCorp, with overbilling public agencies hundreds of millions of dollars for blood tests and other diagnostic work.

Fritsch said that Carreyrou's name had come up during his chat with the whistleblower, whose name was Chris Riedel. "I reached out to him and identified myself as a researcher and former WSJ reporter," Fritsch wrote. "To which he says, 'Oh, so do you know Carreyrou?' Sure, says I."

Fritsch then switched gears. "I caught him lying to me about something and just wanted to reach out and get your read of this dude," he added. "Hope you are well and after the bad guys."

Carreyrou thanked Fritsch for the kind words about the Pulitzer but said he liked Riedel. "I do know Riedel and what I can tell you is he hasn't steered me wrong on anything I've discussed with him," he wrote. "I haven't experienced any lying by him. That said, he's clearly a self-interested guy, but aren't we all?"

Carreyrou didn't sense there was anything unusual about Fritsch's email or its timing. But as it happened, the reporter had just started his own investigation of the medical testing industry. That probe would focus on a company whose name, thanks to

Carreyrou's dogged efforts, would become synonymous with one of the biggest financial and medical frauds of the past decade—Theranos. Carreyrou didn't know it but Fusion GPS was already working on the company's behalf.

JOHN CARREYROU CHRONICLED HIS Theranos investigation in a 2018 best-selling book, *Bad Blood*. In 2015, when Carreyrou began looking into the company, Theranos and its founder, Elizabeth Holmes, were receiving adoring media attention and the start-up testing company had a market value of $9 billion. Its board of directors included a former secretary of state, George Shultz, and other luminaries.

Theranos, based in Palo Alto, California, claimed to have developed a revolutionary testing technology. Traditionally, a large hypodermic needle is used to draw blood from a patient's arm to get samples for analysis, a sometimes painful process. Holmes said the Theranos system could yield similar test results using only a finger prick's worth of blood. A big U.S. pharmacy chain, Walgreens, had struck a deal with Theranos to place testing kiosks in its drugstores nationwide.

Carreyrou, however, had gotten a tip that Theranos's technology didn't work and was producing misleading data. If true, the device's failure held consequences not just for investors but for doctors and patients making decisions based on those findings. He started secretly meeting with company insiders who told him that Holmes and others were engaged in a brazen cover-up while they tried to fix the device's problems.

In April 2015, Theranos finally caught wind that Carreyrou was poking around and hired an outside crisis management firm, which contacted him. Right around the same time, a law firm run by David Boies, a well-known attorney who represented Theranos, retained Fusion GPS to work on the lab company's behalf.

Five days after his initial email to Carreyrou, Fritsch sent him a follow-up note disclosing Fusion GPS's ties to Theranos. "Hey, so something came up I feel like I should disclose," Fritsch wrote. "So we've been looking at lab companies as part of a litigation review for a law firm which was the context of my email to you. I've since learned that this is related to representation of Theranos and that you have a reporting interest in them. I learned this because I was asked, as a former WSJ guy, do you know Carreyrou? I said: Yeah, he is a solid citizen and just won a big prize etc."

Fritsch added that he was urging Theranos executives to cooperate with Carreyrou's inquiries about the company's technology. "I very strongly encouraged them to be open and fulsome in giving you access and dealing with your questions," he wrote. "Said that it was in their interest because the current canon of coverage of them (imo) borders on hagiography and they need to starting dealing with the serious media fluent in things. . . . Anyway, wanted you to know this and that a) if that means you don't wanna talk further, cool and b), if you do, cool. . . ."

Carreyrou responded, saying that he welcomed Fritsch's offer of help. "Thanks for disclosing that," he wrote. "I'll be frank with you: I'm working on a very serious story about them. Of the variety that I tend to do, which is not in the puff piece genre. I've been working on it for months. It's eye-opening stuff. I can't say more than that, but they'll now [*sic*] what I have soon. I'll be starting to run it by them next week."

Carreyrou and Fritsch began regularly exchanging emails about Theranos, its technology, and Elizabeth Holmes, the firm's founder. Holmes wore black turtlenecks to emulate the style of Apple's leader, Steve Jobs. She kept her blond hair tightly pulled back and had trained herself to look straight ahead for minutes without blinking. She also spoke in a low, husky voice. "Is she a cult leader? On medication? What?" Fritsch remarked in one note to Carreyrou.

In another email, Fritsch said he understood the reporter's skepticism about the accuracy of Theranos's testing device but added that he had confidence in the company because a lawyer he knew named Heather King had just left David Boies's firm to become general counsel of Theranos. Fritsch didn't mention in his note that King had used Fusion GPS and him on an earlier case.

"They just named a new GC, a gal I know very well and think is very smart," Fritsch wrote. "Haven't asked why she jumped from Boies Schiller (where she was a partner) but she wouldn't have done so lightly and without some real due diligence."

Soon, however, Carreyrou started to lose patience. Holmes was cooperating with reporters at *The New Yorker*, *Fortune*, and other publications that were doing favorable pieces about her and she was appearing on television programs for softball interviews with Charlie Rose and other hosts. But Theranos still hadn't replied to a list of questions he sent the company and was only offering to send lawyers and public relations executives to meet with him.

Carreyrou wrote to Fritsch in mid-2015 that the *Journal* would cancel a planned meeting with Theranos representatives if Holmes or the company's president, Ramesh "Sunny" Balwani, didn't attend. Balwani, the reporter's sources had told him, was Holmes's lover and a key player in the cover-up.

"It will have to involve either Elizabeth, Sunny or another company executive," Carreyrou wrote. "Otherwise, it's not worth our while."

Right about then, Fritsch's tone turned from seemingly friendly and helpful to dismissive and condescending. He told the reporter he had some sound advice for him based on the many years he had spent at the *Journal*.

"You are talking to the wrong dude, truly," Fritsch wrote. "I do have some influence and I have exercised it in the interests of both parties because I think the advice is the same: more disclosure and more access is always the way to go. I learned that over 17 years at the *WSJ*. But with all due respect, I think you're playing this a lot harder than it needs to play. I'd simply note that Charlie Rose and their brother didn't start with a series of questions that basically accused Sunny and Elizabeth of fraud and sleeping together.

"I get the tactic and have used it myself but usually only after I had the Abu Ghraib photos [of U.S. soldiers torturing captured prisoners in Iraq] in my hand, so to speak. So you have a bit of work to do to convince people that you are amenable to an alternative narrative."

Carreyrou quickly wrote back. "Hey man, I appreciate your back channeling and attempts to smooth the dialogue between me and the company," he replied. "But what I can certainly do without is patronizing commentary from you about my reporting or approach to reporting. We are both big boys here. I've been doing this for 20 years. If you did things differently when you were a reporter, that's fine. But let's keep it respectful. Otherwise, it will cease being productive."

"Big boys we are," Fritsch replied. "Which is why I'm giving you the respect to not pull punches and tell you what I see. That ain't being patronizing. It's trying to help you understand the company's frame of mind. Look I have nothing but respect for your reporting effort and commitment not to take the guided tour. The truth is my commitment too. So if you've got 'em, I'm gonna be the first one to congratulate you."

A FEW WEEKS LATER, Peter Fritsch came to the *Journal* as part of the Theranos delegation that met with John Carreyrou and editors at the paper. The group did not include Elizabeth Holmes or "Sunny" Balwani. Instead, David Boies and Heather King, the lawyers, were there along with one Theranos executive. As Carreyrou later recounted, King started the meeting by declaring that all the reporters' questions about apparent flaws in the company's technology were based on "false premises."

King then announced that Theranos knew the identity of one of Carreyrou's sources. She said it was a young researcher at the company, Tyler Shultz, who was the grandson of George Shultz, the former secretary of state. The younger Shultz had faced scrutiny from lawyers from Theranos and, several weeks before the *Journal* meeting, Fritsch had made a cryptic remark to Carreyrou about the young researcher that the reporter sensed was a trick to get him to say he was a source. In both cases, Carreyrou didn't take the bait.

As the meeting ended, Fritsch told the *Journal* editors present that he expected them to require Carreyrou to identify his sources in any articles about Theranos, using the same standards imposed when he was at the paper. "I just ask one thing as a former practitioner," he said. "At some point, I hope some of the people who are critics are going to be willing to go on the record as well, so we're not massing negative blind quotes. That used to be a thing for us."

AFTER THE MEETING, JOHN Carreyrou's investigation of Theranos continued. Fritsch focused on a different kind of investigation, one about his former colleague. He wanted to know what Carreyrou had learned from public officials about Theranos.

Federal and state agencies are required by law to release certain types of information to journalists and other members of the public who make requests under so-called open records laws. When media organizations make such requests, they typically encourage staffers to acknowledge their journalistic affiliations. By doing so, reporters can't be accused afterward by the subjects of an investigation of going behind their back. But there is a downside to such transparency. Open record requests become public documents and people can learn from them what news organizations are investigating. Private operatives, lawyers, and crisis management experts have little interest in transparency, so they often cloak their open records requests by laundering them through a contractor to make it appear that an inquiry is coming from someone else.

For its part, Fusion GPS had long employed a former journalist who was an expert on open records laws as a contractor. That ex-reporter, Russell Carollo, had spent many years at the *Dayton Daily News*, where he had won numerous awards including a Pulitzer Prize for articles exposing medical malpractice by military doctors. He had a reputation for being cantankerous and once told a colleague that he didn't just burn his bridges, he "blew them up."

By the late 2000s, Carollo had run through a series of newspapers and was having difficulty finding a job. He faced another challenge. He was diagnosed with early symptoms of Parkinson's disease. Carollo retired and purchased an isolated home about thirty minutes from Pueblo, Colorado, a scenic town two hours south of Denver. Inside his large, log-cabin-style house, he made a living doing open records requests for a variety of clients.

Carollo first started working for Simpson back in 2009, during the days of SNS Global. By 2015, he was employed on a nearly full-time basis by Fusion GPS, and that August, about two months after John Carreyrou's meeting at the *Journal* with Peter Fritsch and Theranos representatives, he began preparing public records requests at Fritsch's direction that were intended to uncover the types of documents that Carreyrou had sought about Theranos.

Before sending out one request, Carollo sent Fritsch a draft of it. The inquiry was seeking the names of individuals who had asked regulators for copies of inspections conducted at Theranos. Those inspections, known as "2567" reports, list deficiencies and other problems found at a company's laboratories.

In the draft, Carollo included requests for reports made by Carreyrou, and Fritsch apparently realized that if the reporter saw it he would realize that Fusion GPS was using Carollo as a cutout to monitor him.

"I would like to not mention Carreyrou by name," Fritsch wrote Carollo. "Can we embed him in a broader request that says: 'any requests by media organizations for 2567s made in the past two years' or some such so that it doesn't look like we are targetting [*sic*] him specifically? the reason is obvious: if we name him and he sees that, he'll know who you are working for/with etc. . . ."

Carollo replied that he thought Fritsch's shotgun-style approach was too broad to fly with officials who approved requests. Fritsch had another suggestion. "To mask it, let's also include *The New York Times*," he wrote.

JOHN CARREYROU'S EXPOSÉ OF Theranos appeared on the front page of *The Wall Street Journal* on October 15, 2015, some ten months after his investigation started. Its publication marked the beginning of the end for Theranos. In time, Elizabeth Holmes faced civil and criminal investigations.

On the same day as the article's publication, another contractor working for Fusion GPS or Theranos was still trying to dig up information about Carreyrou and his sources. That contractor sent an open records request to federal agencies seeking copies of all communications between Carreyrou, his editors, or "any other employee, representative, or agent of *The Wall Street Journal*, News

Corp or Dow Jones & Company" with the federal Food and Drug Administration or the Centers for Medicare & Medicaid Services, agencies that regulated Theranos.

Carreyrou only learned in late 2017 that Fritsch had been bird-dogging him when he was contacted by a reporter from *The Washington Post*, who was doing a piece about Fusion GPS. "It was infuriating and despicable," Carreyrou would say later about Fritsch's actions. At the time, he was doing final revisions on *Bad Blood* and he briefly thought about adding a section about the Fusion GPS operative in his book, but he decided it was too late.

CHAPTER 6

UKRAINE TOMORROW

LONDON, 2016

In the summer of 2016, an investigator for Global Witness, the anti-corruption group, sat in a London café listening to a woman describe the documentary she was making. Charlotte Marie introduced herself as a fashion student from Paris who was drawn to filmmaking in order to chronicle the plight of political refugees fleeing the Democratic Republic of Congo to escape its autocratic ruler, Joseph Kabila.

The Global Witness official, Daniel Balint-Kurti, knew a lot about the Congo and Kabila because he had spent nearly a decade as a journalist in Africa, reporting on armed conflicts, atrocities, and corrupt business dealings. Balint-Kurti, who was thin and quiet, had become familiar with every company and businessman exploiting the continent's natural resources. Along with ENRC, the mining company founded by the Kazakh "Trio," that list included Glencore, a Swiss firm founded by Marc Rich, the controversial American businessman pardoned by President Bill Clinton after his conviction for a tax-related crime; Oleg Deripaska, the Russian oligarch; and two Israeli businessmen, Dan Gertler and Beny Steinmetz. As head of investigations at Global Witness, Balint-Kurti oversaw the group's inquiries, including one that was examining whether ENRC or Dan Gertler had bribed President Kabila of Congo to acquire mining rights there. (ENRC and Gertler have denied doing so.)

Balint-Kurti was cautious by nature and he usually checked out strangers before agreeing to see them. He hadn't done so with Charlotte Marie, who looked to be in her late twenties, with long dark hair, but he was impressed by her passion as they spoke. She became animated while recounting a meeting with one member of the Congolese diaspora. The man, she said, was working in a supermarket in France and spontaneously announced to her that he planned one day to return to his homeland and run for president.

When Balint-Kurti asked her who was underwriting the film, she named a Swedish philanthropist, Leonard Boden. A few days after their meeting, Balint-Kurti did a Google search for Boden. He didn't find any Swedish philanthropists by that name. Instead, the only "Leonard Boden" of any note who turned up was a deceased British painter known for his portraits of Queen Elizabeth II.

Not long afterward, Marie emailed Balint-Kurti proposing another meeting. This time he didn't reply. Instead, he made a note in his diary describing their encounter and his failure to find her supposed benefactor. "She seemed very nice and her film project seemed interesting," he wrote. "I am, however, wary of such meetings, as I can never exclude the possibility that my interlocutor has been sent to spy on me."

ABOUT A YEAR LATER, Daniel Balint-Kurti got a call from a journalist with an Israeli television show that specialized in investigative exposés. The reporter told him that the show was doing an episode about Black Cube, the Israeli investigative firm, and whether it had been hired to work on behalf of the Congo's leader, Joseph Kabila, and monitor his opponents at home and abroad. (Black Cube would later deny it did any work for the Congolese government.)

Balint-Kurti and the reporter had several discussions before he mentioned his encounter with Charlotte Marie. It proved a conversation stopper. The Israeli journalist told him that he and his colleagues had interviewed another person who said they had been approached by a woman matching Marie's description. She had described herself as a French fashion student making a documentary about the Congolese diaspora and told the same anecdote about meeting an exile who hoped one day to be the Congo's president. The Israeli reporter and his colleagues didn't believe Charlotte Marie was the woman's real name. They didn't say what it was. But they felt certain about one thing—she was a spy for Black Cube.

MOST PEOPLE FIRST HEARD about Black Cube in 2017, when the Harvey Weinstein case erupted. Along with the Israeli firm, the producer and his lawyers also hired K2 Intelligence, Kroll, and the same San Francisco–based firm that had worked in the 1990s for Bill Clinton's campaign, to investigate actresses and other women accusing Weinstein of rape or sexual abuse.

Still, it was Black Cube that grabbed the attention. In news accounts, the firm was depicted as the practitioner of a new and particularly virulent form of private spying with its tentacles reaching everywhere. Spies for Black Cube, according to one published report, were behind a "dirty ops" campaign to collect dirt on former members of the Obama administration who had helped negotiate the Iranian nuclear arms control deal. Black Cube operatives had traveled to Canada, other articles recounted, where they adopted the guise of business executives and tried to bait a legal arbitrator into making anti-Semitic remarks in order to get him dismissed from a case. In Eastern Europe, according to other accounts, the firm's spies had infiltrated prodemocracy groups and wiretapped a law enforcement official.

Whenever its name appeared in the news, Black Cube declined to identify its clients and never confirmed or denied any speculation made with regard to the company's work. The firm also claimed that it applied high moral standards to its work and that major law firms had reviewed and approved the techniques used in its operations and that the firm's methods complied with the laws of the jurisdictions in which it operated.

In truth, Black Cube really wasn't doing anything that new or unique. If anything, it specialized in running the oldest con in the book—sizing up a mark and playing them for a sucker.

Just about every scam devised—get-rich quick deals, lonely hearts proposals, three-card monte games, email schemes of various types—revolve around a basic principle. A sharpie tries to get what they want by having a sucker respond impulsively or emotionally to the bait they toss out. The button that a con artist or a private operative pushes can vary. It might be an overeagerness to make money, receive recognition, find love, get sympathy, or fulfill some other basic human desire. The con works when the target, rather than asking themselves what is going on or why a stranger has suddenly contacted them, reacts without thinking.

BY THE 2010S, OPERATIVES-FOR-HIRE were increasingly manipulating the internet and social media in ways that would benefit a client or damage a target. K2 Intelligence, Kroll, and other large investigative firms offered "reputation management" services to oligarchs and other clients, a euphemism for the process where a person's or a company's online profile was cleansed of negative information. Private spies were also turning social media and internet postings into offensive weapons that they used to surveil targets, covertly trash adversaries, and manipulate public opinion. There were lots of ways to do so. Private spies created Twitter bots to follow people they were tracking or deployed fake internet personas to engage unwitting targets in conversations. They salted citations on Wikipedia with negative information or created fake web pages about a client's opponent.

For Black Cube, digital tricks were a way of life. The firm utilized a cadre of operatives who took on false names and phony identities and then approached targets, hoping to dupe them. For each mission, social media experts employed by Black Cube—referred to as "avatar operators"—provided an operative with an online identity, or "legend," that consisted of a fake name, a phony past, and an imaginary job history. A Facebook page was created under the operative's assumed name that was embellished with the names of their nonexistent friends. A LinkedIn profile listed the operative's supposed employment profile, including phony testimonials about their skills. In case a target got curious and clicked on the link of the company where an operative now supposedly worked, they would be taken to a web page created for the nonexistent firm, illustrated with stock photographs of a corporate office. Black Cube analysts researched an operation's targets, creating profiles of them and their potential vulnerabilities.

One of Black Cube's operatives, Stella Penn Pechanac, was a veteran of the Israeli military and a former aspiring actress who played a variety of roles for the firm. On behalf of Harvey Weinstein, she had introduced herself to one of the producer's accusers, the actress Rose McGowan, as "Diana Filip" and told her she was an executive at an investment fund in London called Reuben Capital Partners that promoted women's rights. Pechanac's assignment as "Diana Filip" was to get McGowan to give her parts of a forthcoming Hollywood memoir she had written that mentioned Weinstein. She spent months grooming McGowan by showering her with attention and sympathizing with her concerns about how Hollywood treated women. For another assignment, Pechanac said her name was "Maja Lazarova" and that she worked for an executive recruitment firm in London called Caesar & Company. On yet another case, she called herself "Vanessa Collins."

"Tame" journalists got scripts about how to approach targets in order to trick them into thinking they were being interviewed by real reporters. One apparent target was a controversial Romanian businessman named Frank Timis. An assignment sheet headed "Cover Story" read: "You have come across Frank Timis' name following the political scandal cases by his project in Senegal. You have found that Frank's personal history is an inspiring story of a person who became wealthy and successful against all the odds. You offer Frank a personal interview in which he will describe his inspirational life story, focusing on how he made a fortune." That was the bait. The real objective was different. "The main target of the project, in addition to gather incriminating information about Frank, is to find out [what happened to] the large amounts of money that were invested" in one of his projects, according to a document formatted in a style that Black Cube often used.

Black Cube had a few ground rules intended to keep itself out of legal trouble. The false identities adopted by operatives weren't supposed to resemble anyone real. But apart from some restrictions, operatives faced few limits. "In order to get people to talk you need to create a specific world which is a virtual world," an advisor to Black Cube told *The Wall Street Journal*. "It's like creating a play."

BLACK CUBE OPENED IN 2010 right about the same time that Glenn Simpson and Christopher Steele went into business. Its first big case foreshadowed what lay ahead. The firm was hired by two real estate developers in London facing investigation by Britain's Serious Fraud Office for allegedly defrauding Iceland's largest bank and causing its collapse.

Black Cube operatives found evidence that undercut the agency's case and British authorities were forced to pay the developers millions of dollars in compensation for their legal costs. But as soon as the case was over, the developers and Black Cube executives were at each other's throats over how much the Israeli firm was owed. One of the businessmen reportedly acknowledged that he had secretly recorded the firm's operatives and Black Cube claimed the developers had reneged on a deal that would give the firm a cut of any money they were spared from paying out. On the eve of a critical hearing, the dispute was settled.

OVER TIME, BLACK CUBE would grow to employ more than one hundred people, and cloak itself in the mystique of the Mossad, Israel's feared intelligence agency. One of its directors was the former head of the spy agency and Black Cube boasted in marketing pitches that staffers were "highly experienced and trained in Israel's elite military and government intelligence units." Much of Black Cube's work involved run-of-the-mill business disputes in Israel, Europe, and elsewhere. But the firm had interesting clients including Harvey Weinstein and operatives associated with the firm apparently tried to land other controversial customers, including associates of a Ukrainian oligarch who would later emerge as a player in the impeachment proceedings brought against President Donald Trump. The oligarch, Dmitry Firtash, also happened to have once headed the Ukrainian pipeline company that Glenn Simpson and Global Witness suspected of being secretly controlled by a top Russian gangster.

In 2014, Firtash was indicted by the U.S. Justice Department on bribery charges, though the allegations were unrelated to the energy business in Ukraine. Instead, federal prosecutors charged that he and others had paid more than $18 million in bribes to officials in India to secure rights there to mine titanium, a valuable mineral. Firtash was arrested in Austria and freed from custody after he posted $174 million in bail and agreed not to leave the country while he fought extradition to the United States.

Firtash, fresh from his release from jail, spoke at a 2015 business conference in Vienna called "Ukraine Tomorrow." Some participants in the event apparently contacted Black Cube operatives who, without alerting the company, then devised a business pitch intended to generate publicity painting Firtash as a victim of Obama administration policies. The proposal, which was dummied-up to resemble a Black Cube memo, was titled "Client's Narrative" and stated, among other things, that, "Via a plethora of 'lawfare' tactics (RICO, sanctions, etc.) the U.S. have attacked Firtash and his allies, removing them from positions of influence in order to make way for the US 'puppet government'" in Ukraine. It listed specific steps that Obama administration officials and American lawmakers had supposedly taken against Firtash and other Ukrainian oligarchs with ties to Moscow. "There are several businessmen [in Ukraine] that U.S. does not act against, that it should," the memo stated. "Therefore, the claim that Firtash is being wanted by U.S. for corruption schemes is at best double standards."

The pitch didn't generate any business or favorable publicity for the oligarch. In 2019, Firtash remained in Austria fighting extradition. He told *The New York Times* that during the Ukraine-related controversy that engulfed President Trump he was approached by an intermediary working with former New York City mayor Rudolph Giuliani, a top Trump defender, who was seeking dirt on former vice president Joe Biden.

BLACK CUBE DID WORK on behalf of oligarchs and despots in Eastern Europe, including allies of Hungary's autocratic ruler, Viktor Orban. In 2018, prodemocracy and other advocacy groups in Hungary campaigning for greater political and social freedoms started receiving emails that were supposedly from executives of companies who said they were eager to meet with the activists and donate funds to their organizations. Most activists didn't respond. But those who did were invited to meetings at fancy restaurants in Budapest, Vienna, and Paris, where, after some small talk, Black Cube operatives posing as executives peppered them with questions about George Soros, the Hungarian-born investor who supported prodemocracy efforts.

Authoritarian leaders throughout Eastern Europe had long painted Soros as the leader of a global cabal that used activists as proxies in a quest to seize power. Tape recordings secretly made during the meetings between activists and Black Cube operatives were edited by an unknown third party to create the impression the activists were in Soros's pocket and leaked to government-friendly media outlets.

"If Soros' people have influence in government, they will occupy the Hungarian energy sector and the banking system," Viktor Orban declared after one report was published. "The Hungarian people will pay a price for that."

BLACK CUBE, HOWEVER, HAD one big problem. It wasn't good at what it did. The firm, despite the high prices it charged clients, kept recycling tactics from one case to another one. The result was that some of its operations looked like bungling, low-rent clown shows. In the case of Hungary, for instance, a female operative who approached activists called herself "Anna Bauer" and said she worked for "Tauro Capital," a London-based company, while other operatives said they worked at another London firm called "Orion Venture Capital." According to a report in *Politico*, the firms' addresses all traced back to a London company that leased out office space and that Black Cube was using as a mail drop. Black Cube, despite its macho marketing image, took the coward's way out whenever its tactics were exposed. It vaporized the fake online identities of its operatives and the websites of phony companies.

AFTER THE FIRM'S WORK in the Harvey Weinstein case came to light, two Black Cube spies decided to seek publicity for themselves. Stella Penn Pechanac, the operative who had approached the actress Rose McGowan, had a compelling backstory. She was born in war-torn Sarajevo and as a child had watched a wounded neighbor bleed to death on her bed. Her family had found safety by moving to Israel, where she served in the military and then attended drama school in the hope of becoming an actor, though that career never took off.

In an interview, Pechanac said she didn't feel she had betrayed other women by spying on them for Weinstein, because her assignment was to find out whether rival Hollywood studios were behind the assault allegations against him. Pechanac, who was tall and blond, was also asked if she had ever used her femininity as a weapon.

"I was never a Bond girl," she replied. "I was James."

A contractor for Black Cube, Seth Freedman, a freelance journalist who employed that role when he contacted women accusing Weinstein of assault, also insisted he hadn't done anything wrong. "I don't feel guilty about anything I did for Black Cube," he told the BBC. "I don't need to sit and stare down the camera and say, 'I'm so guilty; I'm so sorry'—because I'm not."

For its part, Black Cube always insisted it played by the rules. And yet still two Black Cube operatives were convicted in Romania on charges that they had tried to hack into the email of that country's top prosecutor. They were only released a year later from prison when the Israeli government intervened on their behalf.

PLENTY OF EXECUTIVES IN the corporate intelligence industry despised Black Cube. They said they never misrepresented themselves or lied to get information, adding that they viewed Black

Cube's tactics as sleazy, if not illegal. But what really fried them was what happened when they pitched their services to lawyers or other potential clients. Often they were asked if they could do the kinds of things that Black Cube did.

Those requests were hardly surprising. Black Cube and similar firms thrived because their customers—lawyers, companies, powerful individuals, etc.—wanted to win and they often weren't concerned about how that happened, particularly if it was legal and they thought that no one would ever find out about it.

That may be the best, and perhaps only way, to understand why David Boies's firm hitched its star and reputation to Black Cube when it agreed to hire the firm to work on behalf of Harvey Weinstein. The contract signed by Boies's firm was filled with red flags. It mentioned Black Cube's use of "avatar operators" and described how the firm would deploy operatives under fake identities to target his accusers. If that wasn't enough, Black Cube, under that contract, stood to make a $300,000 bonus—a "success fee" it was called—if it succeeded in preventing *The New York Times* from publishing an exposé about Weinstein.

Boies, a renowned superlawyer who prided himself on his reputation as a champion of free speech and same-sex marriage, claimed he never saw the Black Cube deal. But when Ronan Farrow of *The New Yorker* disclosed it, the fallout was swift. At the time Black Cube stood to make a windfall bonus by derailing the *Times* article, Boies also was representing the newspaper in an unrelated libel case.

To its credit, the *Times* didn't mess around. In a public and humiliating move, it promptly fired Boies. "We never contemplated that [his] law firm would contract with an intelligence firm to conduct a secret spying operation aimed at our reporting and our reporters," the paper said in a statement. "Such an operation is reprehensible."

Even after the Black Cube contract became public, Boies, who had also represented Theranos expressed regret but still didn't seem to get it. When asked whether he thought it was a misrepresentation for a Black Cube operative to use a phony name and a fake identity while approaching one of Weinstein's accusers, he split hairs.

"I think it may depend on how significant the misrepresentation is to the person receiving" it, he said.

CHAPTER 7

TABLE NUMBER 6

NEW YORK, 2016

The first time that Glenn Simpson heard Bill Browder's name mentioned in the same breath as Prevezon Holdings, the Russian-owned real estate company, was when he got a call from his old friends at BakerHostetler, the law firm. After it was hired to represent Prevezon, the firm dispatched John Moscow and Mark Cymrot to meet with Prevezon representatives and the company's Russian attorney.

Browder, in his telling of Sergei Magnitsky's story, had depicted himself as a capitalist turned crusader after his colleague's senseless death and had identified Prevezon as a beneficiary of the brazen fraud carried out against Hermitage Capital. Prevezon's lawyer, Natalia Veselnitskaya, told a very different tale, one that Glenn Simpson would amplify in coming years on behalf of Prevezon and its owner. Veselnitskaya claimed that Browder, far from being a hero, was a con man peddling fake tales about Hermitage Capital and Magnitsky's death in order to save his own skin.

"Browder has developed a Hollywood stereotype," Veselnitskaya liked to say. "Why on Earth should Americans buy into the story of that swindler?"

IN TIME, NATALIA VESELNITSKAYA would join the ranks of Russians who achieved notoriety during the 2016 presidential campaign, thanks to her meeting with Donald Trump Jr. at Trump Tower. But back in 2013, she had yet to enter the arena of U.S. politics and, as Prevezon's lawyer, was overseeing a legal and public relations campaign against Bill Browder.

Veselnitskaya insisted that Sergei Magnitsky's death, while unfortunate, wasn't the result of torture or medical neglect but the incompetence of the Russian doctors treating him. And Browder's charges against Prevezon, she added, were fake. She explained that the money-laundering allegations against Prevezon's owner, Denis Katsyv, had been first made by a Russian criminal as part of a shakedown scheme and that Browder had picked up those claims and run with them.

THE PREVEZON CASE WOULD unfold over the course of four years. Inside a Manhattan courthouse, prosecutors argued that the firm had laundered some of the $230 million stolen from Hermitage Capital. John Moscow and Mark Cymrot insisted that Browder had given federal authorities false information about the company and that it had done nothing wrong.

During that time, Glenn Simpson would assist Prevezon both inside and outside the courtroom, often by digging up documents or disparaging Browder. He traced corporate filings and identified a network of offshore companies set up by Hermitage Capital to hold stock it owned in Russian companies and uncovered the names of other investors involved with those entities. He tried to plant negative articles about Bill Browder in the media. And he worked alongside Rinat Akhmetshin and a former *Wall Street Journal* reporter turned lobbyist, Chris Cooper, to publicize a documentary that depicted Browder as a fraud.

AS THE PREVEZON CASE unfolded, it would become increasingly hard to believe that Glenn Simpson and Bill Browder ever exchanged a kind word. The loathing they developed for each other became epic in scale and seemingly Freudian in nature, that special kind of hatred that festers between those blind to their similarities. Simpson liked to suggest that he and Christopher Steele were alike because they were the same age and had been born just a month apart. Using that same yardstick, Simpson and Browder were a much closer match because they were born a day apart.

Superficially, the two men didn't resemble one another. Simpson smiled a lot and had a goofy, disarming manner. Browder was stern and tightly wound. But both of them were relentless self-promoters who were ambitious, obsessive, and controlling. In addition, neither of them forgot a slight.

Browder, like Simpson, was skilled in dealing with journalists. Back in Moscow, he would summon reporters to the offices of Hermitage Capital for what he called "presentations." There were meticulous PowerPoint displays detailing corruption or mismanagement at one of the many Russian companies in which Hermitage Capital had invested funds. The objective was to get a reporter to write about that company's problems and force changes that would boost the value of Hermitage's stake in it. One journalist who knew Simpson and Browder said he warned Simpson that the investor would fight back if he tangled with him. But to Simpson that might have sounded more like an invitation than a warning.

ONE OF THE JOBS that Glenn Simpson took on for Prevezon's lawyers was to physically track down Bill Browder. The attorneys wanted to force him to testify under oath in the Prevezon case but the investor apparently didn't want to get grilled and avoided getting served with a subpoena. In 2014, Simpson noticed that Browder was scheduled to speak at a conference in Aspen, Colorado, and a process server was dispatched to intercept him after the event. But when Browder was handed the subpoena, he dropped it on the ground and ran off.

The next time, things weren't left to chance. In early 2015, Browder appeared on *The Daily Show*, the late-night program hosted by comedian Jon Stewart, to promote his book *Red Notice* and discuss the Magnitsky Act. While the episode's taping was under way, a team of private investigators assembled outside the show's Manhattan studio. Two of the investigators, a man and a woman, loitered near the studio's rear door where guests exited the building. Another private eye sat inside a rental truck that was occupying the space where black cars usually parked to wait for the show's guests.

When Browder left the studio, he saw that his car was parked up the block. As he was walking toward it, an investigator tried to serve him but Browder scrambled away, pushing people and jumping in his car. But it was blocked in and Browder leapt out of the vehicle and took off on foot down a snow-covered street. The whole scene was captured on videotape by one of the private eyes and later posted on YouTube. A friend of Simpson's said he received an email from him with a link to the video. He recalled that Simpson's message declared, "I did this."

SIMPSON ALSO PITCHED JOURNALISTS at *The Wall Street Journal* and other outlets about doing an exposé of Browder, pointing to what he said were inconsistencies in the investor's story about the Hermitage case and describing him as someone who put convenience above conviction. An American by birth, Browder had renounced his U.S. citizenship in 1988 for a British one, apparently so he could pay lower taxes, and he had made gobs of money in Russia at a time when corruption there was rampant. In addition, he had once praised Vladimir Putin and said that he really didn't know Sergei Magnitsky, having seen him only a few times.

During the 2013 trial in Moscow at which Browder was convicted in absentia of tax fraud, Russian prosecutors claimed that the investor had made a fortune through illegal stock trades and various scams, including claiming he employed disabled workers to reap tax breaks. "Glenn tried to sell me on doing a piece about how Browder's story was all a fantasy and a scam," one journalist at *The Wall Street Journal* recalled.

Simpson didn't have much success. *The New Republic* magazine ran an article titled "Fighting Putin Doesn't Make You a Saint." But most reporters weren't interested in spending months investigating Browder and it wasn't because they were enamored of him. Some journalists considered him a smug pain in the ass who was fast to complain to their editors about articles he didn't like and, on occasion, threaten legal action.

The problem for Simpson was that most journalists didn't care whether Browder had cheated on his Russian taxes. And his misdeeds, even if some were true, paled against the outrages in Russia that Simpson had focused on when he was a reporter—the rampant corruption, political repression, and state-sponsored assassinations that were the hallmarks of Vladimir Putin's reign.

FUSION GPS MADE A fortune off the Prevezon case. The total amount is not known but records that later emerged showed the firm billed BakerHostetler some $530,000 in fees and expenses during the first six months of 2016 alone, a period that represented just a fraction of the time it worked on the real estate firm's behalf. Glenn Simpson would later say that Fusion GPS only took on the case because BakerHostetler was one of his oldest and best customers, adding that the law firm was "very conservative" about the clients it accepted. Still, Prevezon was not just any real estate company. The father of its owner, Denis Katsyv, was Pyotr Katsyv, a Putin ally and the vice president of Russian Railways, a huge state-owned business. In addition, the Katsyvs had faced earlier legal problems and another company they controlled settled money-laundering charges in 2010 brought by prosecutors in Israel.

Simpson's obsessive pursuit of Bill Browder may have blinded him to something else—that in working on behalf of Prevezon he had been drawn into what Senate investigators would later describe as a Kremlin-sanctioned influence operation to undercut the Magnitsky Act. Ever since the law's passage, Vladimir Putin had been gunning for Bill Browder's head and if it got lopped off as a result of the Prevezon case, it would save him the effort.

IF THERE WAS ONE person who appeared to understand the intersection between Vladimir Putin's agenda and Prevezon's lawsuit, it was Natalia Veselnitskaya, the company's lawyer. Veselnitskaya, who was in her forties and had dark hair and a round face, wasn't a high-profile attorney in Russia. She was a former prosecutor who practiced in a region called Moscow Oblast, a sprawling area that forms a suburban ring around the Russian city. On a scale of Manhattan equivalents, that made her a lawyer from New Jersey.

Prior to the Prevezon case, Veselnitskaya had never been to

the United States. And like many people who first encounter New York as adults, she fell in love with it. She stayed at five-star hotels near the elegant Rockefeller Center suites of BakerHostetler and became a creature of habit. One morning, she passed a store with a long line of people waiting outside. It was a well-known East Side bakery called Ess-A-Bagel and she started getting a bagel there every day before work. Most nights, she ate dinner at the same restaurant on Madison Avenue, Nello, known for mediocre Italian food and high prices. She ordered the same dish, cannelloni, which costs forty-four dollars, and sat at the same sidewalk table, referred to by the restaurant's staff as Table Number Six, even dining there on cold days under a heat lamp.

Her daily routine took her along Fifth Avenue and past Trump Tower. She stopped sometimes at the Gucci store to shop and bought baseball caps and hand towels at a nearby store to bring home as presents for her family and friends. Glenn Simpson and everyone else who met Veselnitskaya came away with the similar impression. The attorney, a divorced mother of four, was extremely ambitious and she saw the Prevezon case as the springboard to a new high-profile legal career.

IN EARLY 2016, VESELNITSKAYA launched another offensive against Bill Browder and the Magnitsky Act in which Glenn Simpson would take part. It involved a newly formed lobbying group with the tongue-twisting name of the Human Rights Account-ability Global Initiative Foundation, a group that received some $500,000 in funding from Pyotr Katsyv, the Kremlin-connected railway executive and father of Prevezon's owner. On paper, the stated purpose of the organization was to advocate for the ability of Americans to again adopt Russian infants, the program Vladi-mir Putin ended in retaliation for passage of the Magnitsky Act. It was a feint because everyone knew that the Magnitsky Act would have to be watered down or repealed before Putin would allow adoptions to restart.

The website of the Human Rights Accountability Global Initiative Foundation featured photographs of smiling parents and happy children. It described how ending the adoption program had caused the deaths of Russian infants. "Some of the blocked adoptions were taken up by Russian families—while other children were adopted overseas," the website stated. "But some children got sick and others reportedly died."

One American woman who was eager to adopt a Russian child explained to Bloomberg News that she contacted the organization in 2016 after seeing its website. She was told to meet one of its representatives at a sandwich shop inside a Washington train station. When she arrived, she was greeted by Rinat Akhmetshin, who said that he hoped the Russian adoption picture "would change" after the 2016 U.S. presidential election, then not far away.

The creation of the lobbying group occurred just as an expanded version of the Magnitsky Act was working its way through Congress. The new law, named the Global Magnitsky Act, would allow the United States to impose sanctions against individuals involved in human rights abuses in any part of the world, not just Russia.

Lobbyists paid by the Human Rights Accountability Global Initiative Foundation were working Capitol Hill to convince lawyers to drop Magnitsky's name from the new bill and examine whether the passage of the Magnitsky Act was warranted. Later, during congressional testimony, Simpson would say he understood those lobbying efforts were only aimed at the new law and weren't intended to undercut the original Magnitsky Act. But that sounded convenient given that stripping the dead man's name from the new bill would raise questions about the original Magnitsky Act and efforts by Browder to get countries to adopt similar measures.

IN THE SPRING OF 2016, the Newseum, an exhibition hall in Washington, D.C., that celebrated the newspaper industry and free speech, announced the forthcoming screening of a movie made by a Russian documentary filmmaker. The documentary was titled *The Magnitsky Case—Behind the Scenes*, and it presented a view of Bill Browder in line with the one that Natalia Veselnitskaya and Glenn Simpson were promoting.

The filmmaker behind *The Magnitsky Case*, Andrei Nekrasov, had made an earlier, well-regarded documentary about the 2006 death in London of a former KGB spy and Putin critic, Alexander Litvinenko, from plutonium poisoning. That film pointed the finger of blame for Litvinenko's death at the Russian leader and Nekrasov said in interviews that he was initially sympathetic to Browder's story. He said his view changed as he learned more about the case and compared records about Magnitsky's death in their original Russian against the English translations made by Browder's team. "I can prove in court that Browder is not telling the truth," he told one reporter.

The film was a curious mélange of documentary footage and staged reenactments. Veselnitskaya helped arrange screenings of the documentary at several film festivals in Europe but Browder aided by members of Magnitsky's family managed to block the showings. In the United States, Rinat Akhmetshin and Chris Cooper, the former *Wall Street Journal* reporter turned lobbyist, outmaneuvered him. Cooper approached the Newseum about renting out the facility's auditorium for a screening. Once the deal was struck, Simpson started calling journalists he knew in Washington and urged them to see the movie. When Browder caught wind of the plan, he tried to stop it but Newseum officials refused to back down.

THE FILM'S SCREENING WAS scheduled for June 13, 2016. But in the week that preceded it, a cascade of events unfolded that would change the fates of Glenn Simpson, Christopher Steele, Natalia Veselnitskaya, and Rinat Akhmetshin.

Five days before the screening, on June 8, Veselnitskaya arrived in New York on a flight from Moscow. Her agenda was twofold. There was an important hearing the following day in the Prevezon case. But she also planned to use her visit to the United States to undermine the Magnitsky Act and she received an email soon after her arrival confirming a meeting for her on the following day, June 9, at Trump Tower with Donald Trump Jr.

The email was from Rob Goldstone, a publicist for a Russian rock musician, Emin Agalarov, whose father, Aras Agalarov, was the Russian developer who partnered with Donald Trump to bring the 2013 Miss Universe pageant to Moscow. Veselnitskaya told the Agalarovs, who were Putin allies, about her campaign against the Magnitsky Act and, at their behest, Goldstone reached out to Donald Trump Jr., thinking his father would be a natural ally. To stir his interest in meeting Veselnitskaya, Goldstone sent the candidate's son emails, saying there were "some official documents and information that would incriminate Hillary and her dealings with Russia," and that Veselnitskaya was a "Russian government attorney." Donald Trump Jr., in one of the more memorable lines he would ever utter, responded, "If it's what you say, I love it."

On the evening of June 8, Veselnitskaya had dinner with lawyers from BakerHostetler to discuss the next day's hearing, which involved a last-ditch effort by Bill Browder to have John Moscow removed as Prevezon's lawyer. Years earlier, Browder had hired

Moscow to try to find out what happened to the money stolen from Hermitage Capital. Then, the lawyer had gotten involved in a big case and Browder said that he rarely heard from Moscow again until he reemerged in 2013 as a lawyer for Prevezon. Browder claimed that he had shared information with Moscow and that the lawyer now had a conflict of interest. But the judge overseeing the case ruled that Moscow didn't have a conflict because it was the Justice Department, not Browder, that was suing Prevezon.

An appeals panel had agreed to review the issue and, on the morning of June 9, the hearing began. Glenn Simpson came up from Washington to attend the session and he and Veselnitskaya sat in the courtroom. Over dinner, BakerHostetler lawyers had predicted the appeal would be an easy win and the firm brought in one of its biggest guns, former U.S. attorney general Michael Mukasey, to argue that Moscow should be allowed to continue. Veselnitskaya said her limited English prevented her from understanding everything at the hearing but she sensed right away that it wasn't going well. The judges looked irritated by Mukasey's performance, which she described as arrogant and overconfident. "I knew right then we were going to lose," she remarked long afterward. "I have devoted my adult life to this profession and it doesn't matter what language is spoken in the courtroom, there is a professional understanding of what is going on and what is coming."

Months later, when the panel ruled, her prediction proved true. Moscow was barred from the case and Veselnitskaya had to find another U.S. law firm to represent Prevezon. But on June 9, there were more pressing matters ahead at Trump Tower. After Simpson left the courthouse, she went to Nello, where she met a business associate of Aras Agalarov, the developer. She also called Rinat Akhmetshin, who, by chance, happened to be in New York that day. She asked him to join her at Nello and told him about the Trump Tower meeting. Following lunch, they walked to the building and rode an elevator up to the twenty-fifth floor, where they entered through a set of double doors into a reception area. Veselnitskaya and Akhmetshin were ushered into a set of suites where Donald Trump Jr. has his office. Along with Trump, Jared Kushner, his brother-in-law, and Paul Manafort, the former lobbyist who was now managing Donald Trump's presidential campaign, were waiting for them.

After the meeting, Veselnitskaya and Donald Trump Jr. seemed to agree that Rob Goldstone, to hype their interest and expectations, had sold them a bill of goods. Donald Trump Jr. expected dirt on Hillary Clinton. Instead he got a memo from Veselnitskaya that Simpson had prepared for use in the Prevezon case about some American investors who were connected to Browder's business dealings in Russia and were Clinton donors. Veselnitskaya said she expected to meet with Trump lawyers who could juice her lobbying campaign against the Magnitsky Act. Instead, she got the candidate's son, who struck her as a clueless, entitled rich boy.

When Veselnitskaya started giving him her pitch about Browder, the Magnitsky Act, and Russian adoptions, Trump's eyes glazed over and Kushner left the room. Less than thirty minutes later, the meeting was over. It was a flop.

ON THE NEXT DAY, June 10, Akhmetshin and Veselnitskaya took the train to Washington, D.C. The screening of *The Magnitsky Case* at the Newseum was now three days off and they were eager to step up their lobbying activities for it. That evening, a group dinner was held at a Washington restaurant, the Barcelona. Along with the two Russians, Glenn Simpson, his wife, Mary Jacoby, and Mark Cymrot, the BakerHostetler lawyer, attended it.

There were two last-minute guests as well, a Washington-area couple, Marie Arana and Jonathan Yardley. Both were authors and former literary critics at *The Washington Post* who lived just a few doors away from Akhmetshin but barely knew him and had no idea what he did for a living. He would wave to them as he pedalled by, transporting his daughter to school on the back of his orange bicycle. He was always friendly and when he learned the couple was traveling abroad, he would offer them restaurant recommendations in cities on their itinerary.

At the restaurant, Arana was seated next to Simpson and they spoke about journalism, Latin America, and a biography of Simon Bolívar she had recently written. Later on, she would recall scattered comments that were made about the Magnitsky Act and a documentary but none of it made any sense. When she and her husband returned home, both wondered why they had been invited to the dinner.

The screening of *The Magnitsky Case* would go forward. Veselnitskaya and Akhmetshin would never tell Simpson about their visit to Trump Tower. He would also never tell them about his secret, one that involved a resident of that building. He had just hired Christopher Steele to dig up dirt in Russia on Donald Trump.

CHAPTER 8

GLENNTOURAGE

WASHINGTON, D.C., 2016

Glenn Simpson loved holding court with reporters, regaling them with stories from his newspaper days and presenting himself as a journalistic wise man. Peter Fritsch described his partner's routine to one friend as his "Glenntourage."

In the fall of 2016, Simpson was at a conference of documentary filmmakers and reporters speaking about his work as a private investigator. His presentation was titled "Investigations with an Agenda: Navigating Loaded Terrain" and, as part of the talk, he recounted a recent case. Planned Parenthood had sought his help, he said, after anti-abortion activists released videotapes in which clinic employees appeared to be discussing the commercial sale of tissue harvested from aborted fetuses.

To make the tapes, the pro-life activists, who described themselves as undercover citizen journalists, had posed as employees of a medical tissue collection company and gained entry to Planned Parenthood events, where they secretly recorded conversations. The tapes created an outcry when they were released and Planned Parenthood faced a threatened cutoff of federal funds.

Simpson, along with a video forensics expert and a television producer, examined the tapes to determine if the activists had manipulated them. He challenged the anti-abortion group to release the tape's unedited versions so they could be independently inspected and when it refused to do so, the controversy faded.

"It was a classic example of how people can be manipulative and underhanded under the guise of investigations," Simpson said. "It was fake journalism, because people were posing as journalists and acting as journalists but doing things that journalists never do."

Simpson explained to his audience that he and Peter Fritsch had started Fusion GPS to continue their work as journalists who righted wrongs. "I like to call it journalism for rent, we like to do things in a journalistic way because I think it is a very effective method," he said. "People who have never been a reporter don't understand the challenges of printing what you know, right, because you can't just say what you know, you have to say how you know and you have to prove it. And that sort of imposes a discipline to the investigative process that people in other fields don't really absorb. . . . When you're a spy, you really don't have to get into a lot of that stuff."

HIRED OPERATIVES HAVE ALWAYS spun yarns and self-aggrandized. The first private detective in the United States to gain fame, Allan Pinkerton, employed ghostwriters who cranked out books depicting him as a heroic, civic-minded truth-seeker. Terry Lenzner, the former Watergate staffer who later worked for Big Tobacco, described himself as an "ethical investigator." Jules Kroll liked to say that his firm did "well by doing good," a concept hard to square with spying on public health activists.

Glenn Simpson's motto—"Journalism for Rent"—was happy talk, too. Fusion GPS made big money in the same way that other operatives-for-hire did, by serving powerful people and corporations, aiding investors seeking an edge, and assisting corporations in the crosshairs of prosecutors, regulators, and real journalists.

By 2016, the firm's list of clients included Theranos, the diagnostics company that was accused of fraud; Prevezon, the Russian-owned real estate firm accused of money laundering; and Derwick Associates, the Venezuelan company suspected of paying off officials in Venezuela. The firm also worked for Herbalife, a nutritional supplements company, and hedge funds.

If Simpson had a unique skill, it was the ease with which he gave off the air of still being a journalist. Throughout his career as an operative, he sold himself as a journalist to clients and reporters alike. He socially hobnobbed with reporters and he, Peter Fritsch, or another Fusion GPS operative regularly attended the investigative journalism conference run by Lowell Bergman at the University of California. Reporters who dealt with Simpson never saw him as a hired gun. To them, he was still "Glenn."

THE CONFERENCE OF FILMMAKERS and journalists where Simpson spoke in 2016 took place in October of that year, just a month ahead of the U.S. presidential election. Simpson was asked whether Fusion GPS did political "oppo" work. Though he was speaking to a roomful of reporters, Simpson responded in the way that a corporate executive might when seeking to avoid a reporter's question. He dodged it. "There is not much money in campaign, sort of traditional campaign opposition research type stuff," he remarked. "Sometimes there are big issue campaigns or big, big elections where there is a lot of money thrown around but most of the time, it is a cannibalized, commoditized business."

But for Fusion GPS, the 2016 presidential campaign season was more like a cash cow. While Simpson was publicly discounting his firm's interest in "oppo," reports from Christopher Steele were pouring into his offices. And well before that, a Republican megadonor who had hoped to stop Trump from gaining the party's nomination had paid Fusion GPS to dig up dirt on him.

That donor was a billionaire businessman, Paul Singer, who headed Elliott Management, an investment fund. Singer was a supporter of Florida senator Marco Rubio's presidential ambitions and was underwriting efforts to torpedo Trump's outsider campaign as it was gathering steam. To mask Singer's payments to Fusion GPS, a now commonplace parlor trick was used—the money was disguised by passing it through a middleman. In this case, the intermediary was a conservative nonprofit organization supported by Singer, the Washington Free Beacon Foundation.

Fusion GPS had previously worked for Singer's fund, Elliott Management, which specialized in so-called vulture investing. The term describes a technique where a speculative investor buys distressed debt from holders at a sharp discount and then seeks to profit by trying to get the debt's issuer to buy it back at a higher price. In many instances, the issuer is a government and Elliott Management's most notorious case involved a decade-long campaign it waged against the government of Argentina after it defaulted on billions in debt. After scooping up those bonds at bargain prices, Singer's firm sought to pressure the Argentine government to redeem them at face value. At one point, it convinced a court in Ghana to seize an Argentinian naval vessel docked there so it could sell off the ship and pocket the proceeds. (After international authorities intervened, the vessel was allowed to leave Ghana.)

Elliott Management had an in-house stable of operatives but during the Argentine debt battle it also used Fusion GPS. In 2013, right around the time the Argentine naval vessel was impounded, Russell Carollo, the open records expert, sent requests to the Department of Defense seeking data about U.S. military sales to Argentina in an apparent effort to identify other assets that Singer's fund might appropriate.

Simpson told a reporter around that same time that Fusion GPS had set up a website filled with negative articles about another hedge fund that had taken a different approach to the Argentine debt issue. The reporter said Simpson bragged that the competing fund would never discover Fusion GPS was behind the website because a developer outside the United States had created it.

IN EARLY 2016, A few months before the anti–Bill Browder film was screened at the Newseum in Washington, D.C., a celebration was held there to commemorate the centennial year of the Pulitzer Prizes. During his career as a journalist, Glenn Simpson had never won a Pulitzer Prize or even been a finalist for one. But that January, he was celebrating another kind of achievement—he had just planted a negative piece about Donald Trump in the news media. The article, published on the website of *Vice News*, dealt with Trump's relationship with Jeffrey Epstein, a money manager and sexual predator who had admitted as part of a plea deal struck in 2007 that he had hired a young teenage girl to give him sexual massages.

The plea deal was back in the news because dozens of other female victims of Epstein were seeking to reopen the case. And with Trump running for the Republican nomination, journalists were interested in his relationship with Epstein. The two men, who socialized together in the early 2000s, both had homes in Florida, and Epstein was a member of the Mar-a-Lago Club, the private resort Trump owned there.

"He is a lot of fun to be with," Trump was quoted as saying in a 2002 *New York* magazine article. "It is even said that he likes beautiful women as much as I do, and many of them are on the younger side. No doubt about it. Jeffrey enjoys his social life."

Simpson was promoting stories to journalists about Trump's ties to Epstein and he found an interested customer in an old friend, Ken Silverstein, a fixture in Washington journalistic circles who was then working as a freelancer. Silverstein said that Simpson sold him on the idea of doing a piece about Trump and Epstein, adding that he had gotten other story ideas over the years from Fusion GPS. "*The New York Times*, I know they work with Fusion," he was quoted as saying. "Fusion works with a lot of big media organizations . . . I have a great relationship with those guys."

MUCH OF THE INITIAL research that Fusion GPS undertook during Trump's run for the Republican presidential nomination involved his convoluted financial and business dealings. Russ Carollo made a wave of record requests to the Federal Trade Commission, the Occupational Safety and Health Administration, and other federal agencies seeking documents about Trump's businesses. Other inquiries went to the state attorney general of New Jersey, for information about casinos the developer operated in Atlantic City.

Carollo also sought records from the FBI about Trump's late father, Fred Trump, who was a real estate developer, and he queried law enforcement agencies for documents about controversial figures with whom Trump had intersected over the years. One of them was an alleged mobster, Robert LiButti, who was photographed alongside Trump during a 1988 WrestleMania event in Atlantic City, New Jersey. Another was one of Trump's personal friends, William Fugazy, a businessman pardoned by President Bill Clinton following his conviction for perjury during a bankruptcy case.

As Trump began to emerge in early 2016 as the likely Republican presidential nominee, reporters intensified their focus on him. That March, Michael Isikoff, Simpson's old friend who was now working for *Yahoo News!*, wrote an article about Robert LiButti. Isikoff quoted LiButti's daughter as saying that her father had gambled millions of dollars at one of Trump's casinos, flown to Atlantic City on Trump's helicopter, and partied on Trump's yacht. When Trump was asked about his relationship with LiButti, he gave the type of response he would give time and time again about his potential ties to unsavory characters. "If he was standing here in front of me, I wouldn't know what he looked like," Trump said.

BY THE SPRING OF 2016, Fusion GPS was generating a series of "oppo" reports that it was distributing to reporters about Trump and his business dealings in Russia and Eastern Europe. Trump's privately held company, the Trump Organization, had long ago stopped building properties and was making money now by licensing the businessman's name to other developers who wanted to put it on hotels and office complexes. Toward that end, Trump and three of his adult children—Donald Jr., Eric, and Ivanka—were aggressively pushing licensing deals in Eastern Europe, Central America, South America, and Canada. And while the prospective business partners involved in those projects were often oligarchs or people suspected of corruption, the Trump Organization didn't seem to conduct due diligence investigations into them.

For decades, Trump had also dreamed about having a Trump Tower built in Moscow. In the mid-1990s, he traveled to the Russian city with his then-wife, Marla Maples, to promote the idea of having a series of Trump-branded apartment buildings built there. "If things go the way Donald Trump plans—and they usually do—he will be the first big-time developer since Josef Stalin to attach his name to high-rise towers in Moscow," one article declared. But things didn't work out and his next big effort came in 2013, when he held the Miss Universe Pageant in Moscow as part of a campaign to win Vladimir Putin's support for a project. "Do you think Putin will be going to The Miss Universe Pageant in November in Moscow—if so, will he become my new best friend," Trump tweeted before that event.

THE "OPPO" MEMOS ABOUT Trump's overseas activities that Fusion GPS distributed to reporters were typically four to six pages long. The reports consisted of so-called open-source information or material drawn from publicly available sources such as newspaper articles or research papers that were then stitched together in narratives and seasoned with speculation.

One report, titled "Trump in Russia," contained highlighted sections describing how "Trump's relationship with Russia goes back to the 1980's" and included a list of "Trump's Statements on Russian Politics [that] are favorable to Putin." Another report was titled "Trump in Georgia" and yet another was headed "Trump's Business Partners in the Trump Soho," a hotel and condo building located in lower Manhattan.

But for Glenn Simpson, Paul Manafort's selection as Donald Trump's convention manager in March 2017 was a red letter day. Simpson had been collecting information on Manafort since his days as a lobbyist in Ukraine and, like many journalists, he believed the political operative was a corrupt sleazeball. He began steering reporters covering the Trump campaign toward Manafort, and Fusion GPS generated at least two reports about him. One of them read like a rambling amalgam of rumors and unconfirmed allegations that cried out for the hands of an editor.

"Concurrent with the ongoing US election, Manafort is also

reportedly intervening in a money-laundering investigation in Liechtenstein," a part of one memo stated. "In response to government inquiries, Manafort and several of his agents have sent harassing letters to investigators. It's unclear if the investigation targets Manafort or his Ukrainian and Russian clients. The investigation could stem from pending Liechtenstein orders freezing funds of former Ukrainian president Yanukovich, who Manafort advised for nearly a decade. This suggests Manafort may continue to lobby on behalf of the disgraced Yanukovich, or perhaps it directly implicates him in the laundering of assets from the Ukrainian state."

In April 2016, Glenn Simpson and Peter Fritsch sat in the Washington, D.C., offices of a law firm, Perkins Coie, meeting with an attorney. By then it was clear that Donald Trump would be Hillary Clinton's opponent and Paul Singer, the vulture investor who had been bankrolling Fusion GPS's oppo research, had called it quits. Marc Elias, the lawyer meeting with Simpson and Fritsch, worked both for the Democratic National Committee and Hillary Clinton's 2016 presidential campaign. And as the Fusion GPS operatives would write in their book, *Crime in Progress*, they told him that they wanted to continue gathering oppo material on Trump, particularly about his activities in Russia. Their pitch struck a chord with Elias. During Clinton's tenure as secretary of state in the Obama administration, she had taken a hostile stance toward Vladimir Putin. Now any information linking Trump to Moscow could provide her with ammunition.

Glenn Simpson never liked Hillary Clinton or her husband. Donald Trump was a different story. Like many reporters who had followed the developer's long career, he viewed Trump as a con man, an inveterate liar, and a publicity-seeking fabulist. If mischievous and malign spirits had gathered to script the ultimate "oppo" shit show they could not have done worse than to cast Trump and

Simpson to star in it. They were made for each other. Trump was prone to conspiracy theories. So was Simpson. Through his reporting career, editors had forced Simpson to remain moored to facts. And during much of his time as a private operative, he had relied on tangible evidence—corporate documents, filings with public agencies, lawsuit disclosures, etc. Simpson may have not gotten everything right but at least the information he was drawing on was in black and white.

Now he was moving into the shadowlands, a place where a journalist or an operative had to trust their sources for information because documents didn't exist and it was difficult to confirm what one was being told. It was the realm of "intelligence" and, time and time again, both in journalism and the private spying business, it was a place where things could go really wrong.

GLENN SIMPSON LIKED TO think of himself as an expert on Russia. He had read tons of articles and books about Russian oligarchs, talked with operatives and experts who had worked in the country, and could reel off the names of Russian gang leaders and criminal organizations. All his information, however, was secondhand. He didn't speak Russian, hadn't spent time in Moscow, and didn't have any of his sources there. For his new assignment for the Clinton campaign, he needed someone closer to the action, someone like Christopher Steele.

In May 2016, Simpson flew to London to meet Steele at a restaurant inside Heathrow Airport. Over the years, they had worked together sporadically. Both their firms were too small to afford to have offices abroad, so when Simpson needed work in England he sometimes used Steele, and Steele might do the same when a job required work in the United States.

Since their initial meeting, Steele had gone through a lot personally and professionally. Just as he was starting his new career as a private spy, his first wife had died after a long illness from cirrhosis of the liver. He remarried but for a time he was a single parent with three young children and had to balance the demands of family life with those of his new company, Orbis Business Intelligence.

Orbis had a good reputation. But it was a small fish within London's investigative industry and it was not the first corporate intelligence firm that a lawyer or company would go to when dealing with a Russian-related issue. That honor belonged to Hakluyt, the firm that had dispatched an operative in the early 2000s to infiltrate Greenpeace while posing as a documentary filmmaker. Hakluyt was also considered a coveted landing spot for former MI6 hands and a place where Steele could initially have made far more money as a private spy than he did starting his own business.

Steele was viewed by industry colleagues as someone who knew a lot about oligarchs and crime groups in Russia and Eastern Europe. But he had blind spots as well. One operative who worked with him said it was his impression that Steele had a limited understanding about corporate structures and the intricacies of financial dealings. Other shortcomings would emerge in his dossier about Trump and Russia.

WHEN THE DOSSIER BECAME public, the news media portrayed Christopher Steele as an investigator apart, the rare hired spy who had set commercial interests aside to sound the alarm about the threat posed by Russia to the West. There is no question that he despised Vladimir Putin and he did so for the right reasons. Still, in most other ways, Steele was just another spy-for-hire. Like competitors, he marketed himself on his ex-government credentials, worked on behalf of oligarchs, and made his living from the same types of cases that his London competitors did.

In 2013, for instance, Steele and Mark Hollingsworth, the journalist-operative, worked together on a case involving ENRC, the mining firm owned by the Kazakh Trio. And Hollingsworth told Steele that he had spoken to a source who had access to a large dump of documents from ENRC and the other Trio-connected company, International Mineral Resources.

> 1. He confirmed that there is what he called a "data dump" of thousands of documents, including emails, relating to a range of issues concerning ENRC, IMR and related companies and individuals. He also confirmed that he had seen internal communications and discussions between ENRC and IMR as recently as June of THIS YEAR. He does not have a copy of the data and said that it was with a source in Russia (the source is a Russian national despite living in the USA for the past 12 years). His Russian contact would be willing to provide some sample documents but to have access to the whole archive would involve a substantial fee.

FOR YEARS, CHRISTOPHER STEELE worked for a London lawyer representing Oleg Deripaska, the Russian aluminum magnate who had ranked among Glenn Simpson's favorite reporting targets. And if that wasn't coincidence enough, one of Steele's assignments was to hunt down Paul Manafort's assets so they could be seized in the event the oligarch won a favorable court judgment in a lawsuit he had filed against him.

The dispute had its roots in a business deal that Deripaska and Manafort had struck in the mid-2000s when the lobbyist was advising the president of Ukraine, a Putin ally. Deripaska had agreed to finance an investment fund that Manafort and his partners would use to purchase companies in Eastern Europe. A decade later, the oligarch charged in a 2014 lawsuit that Manafort and his associates hadn't bought any companies with the $18.9 million he had given them and that his money had disappeared.

FOR PRIVATE INVESTIGATORS, CONDUCTING asset searches was standard fare. Christopher Steele's efforts over the years on behalf of Deripaska would go beyond that. The oligarch had a major problem that he had long been trying to fix: the State Department had put him on its shit list because of his suspected ties to Russian organized crime and, as a result, he couldn't get a business visa to the United States.

In 2007, Glenn Simpson detailed in a *Wall Street Journal* article he wrote with Mary Jacoby, his wife, how Deripaska had hired big-name lobbyists to get the prohibition lifted. That hadn't worked and the oligarch soon tried another approach. In the hope of getting a visa as a reward, he agreed in 2010 to help the FBI find Robert Levinson, the American private investigator who had disappeared in Iran, but that search went nowhere.

Soon Steele was drawn into the oligarch's visa quest. In 2015, he got a call from Bruce Ohr, a friend and fellow Russia expert at the Justice Department who was interested in Deripaska. Ohr told Steele that the U.S. government was looking to recruit Russian oligarchs as informants in order to keep closer tabs on Vladimir Putin, and Deripaska, who had cooperated earlier with the FBI in the hope of getting a visa, seemed like a candidate. Steele contacted Deripaska's London lawyer to set up a meeting. For the ex-spy, playing the role of intermediary between the FBI and oligarchs wasn't that unusual. The reason: along with his day job, Steele had been moonlighting since 2013 as a paid confidential source for the FBI.

That arrangement had started a few years after he provided

information to the FBI that the bureau would later use to investigate bribery at soccer's international governing body, FIFA. Over time, he became friendly with Mike Gaeta, an FBI agent based in Europe, and began feeding information to him.

By 2013, Gaeta decided that Steele deserved compensation for his assistance and the result was a formal agreement under which Steele became a paid source. In that role, he brokered meetings between FBI agents and oligarchs or their lawyers. The FBI wanted information from the oligarchs and the latter undoubtedly wanted something in return.

THE GET-TOGETHER BETWEEN DERIPASKA and Bruce Ohr didn't go well. Ohr explained that the United States was hoping to get a better understanding of ties between Putin's government and Russian organized crime. Deripaska wanted a visa but someone had apparently misjudged his appetite for risk. As far as he knew, Deripaska said, there weren't any ties between the Kremlin and Russian gangsters.

CHAPTER 9

THE PEE TAPE

MOSCOW, 2016

Christopher Steele's first memo about Donald Trump's ties to Moscow would endure as the masterpiece of his oeuvre, a rendering of the supposed entanglements between Trump and Vladimir Putin's government that was epic in scope and etched with gripping detail.

While meeting with Steele at Heathrow Airport, Glenn Simpson laid out the scope of his assignment. It was to gather information about possible Kremlin meddling in the 2016 presidential election and determine if Trump or his staffers had business or financial ties to Russia. Initially, Simpson and Steele had agreed to proceed on a temporary basis, with the former British agent reportedly receiving about $30,000 for a month's work. But in June, when Steele sent his first report back to Fusion GPS, Simpson and Peter Fritsch wanted more.

That report was titled "US Presidential Election: Republican Candidate Donald Trump's Activities in Russia and Compromising Relationship with the Kremlin." In it, Steele reported that the Kremlin, according to his sources, had long been seeking to recruit

Trump as an asset by offering him real estate deals. Now Moscow was supplying the Republican presidential candidate and his associates with something even juicier—high-grade political "oppo" that they could deploy against his Democratic opponent, Hillary Clinton. "He and his inner circle have accepted a regular flow of intelligence from the Kremlin, including on his Democratic and other political rivals," Steele wrote.

He added that the FSB, a successor agency to the KGB, was aware of Trump's sexual antics during his earlier trips to Moscow and was ready to use them against him as "kompromat" if he didn't play ball. "Former top Russian Intelligence officer claims FSB has compromised TRUMP through his activities in Moscow sufficiently to be able to blackmail him," Steele wrote. "According to several knowledgeable sources, his conduct in Moscow has included several perverted sexual acts which have been arranged/monitored by the FSB."

It was in that report that the former MI6 agent memorialized what may prove to be his most enduring legacy as a private spy—the tale of the "pee tape." Later on, his partner in Orbis Business Intelligence, Christopher Burrows, would question whether Steele had been wise to pass along the story of how Trump supposedly had hired prostitutes to urinate on a bed used by Barack Obama at the Ritz-Carlton hotel in Moscow. But Steele would insist that he included it because his job was to pass on all the intelligence he gathered rather than "cherry pick" information.

Steele believed that the incident was widely known in Moscow because he cited multiple people as being aware of it. In his memos, he didn't identify his informants or their sources by name but used a common spy world technique—he referred to them by code names, in his case, letters of the alphabet.

Steele wrote that "Source D" had "been present" at the Ritz-Carlton hotel when the incident took place and that two additional people—"Source E" and "Source F"—had confirmed it had happened, either directly or indirectly. Russian intelligence, he was told, had also captured the "golden shower" scene on tape. "The hotel was known to be under FSB control with microphones and concealed cameras in all the main rooms to record anything they wanted to," Steele wrote.

ALONG WITH CHRISTOPHER STEELE, reporters were also scrubbing Donald Trump's past in the run-up to the 2016 election, including rumors about his sexual escapades in Moscow. One of them was Alan Cullison, a veteran foreign correspondent for *The Wall Street Journal* who had spent two decades in Russia. The gruff and plain-spoken Cullison was now based in the Washington, D.C., bureau of the *Journal* but often flew back to Russia or Eastern Europe to cover stories.

When he arrived in Moscow in May 2016 to report about Trump, Cullison started looking into an American-born businessman who was involved decades earlier in Trump's efforts to develop a hotel there. The businessman, David Geovanis, had become a Russian citizen and there were rumors he served as Trump's fixer during his Moscow visits, arranging for him to meet the type of young, Slavic women for whom he had an affinity. A portrait of Geovanis showed him dressed in a tuxedo, standing triumphantly in front of three barely clad and provocatively posed women.

Cullison had known Glenn Simpson since their days together as reporters at the *Journal* and told him about his interest in Geovanis. Cullison didn't think other journalists were looking into the businessman but soon he started hearing the footsteps of other reporters. He suspected that Simpson or Fritsch had leaked his

interest to other journalists and decided not to tell them anything
again.

BY THE SUMMER OF 2016, Glenn Simpson and Peter Fritsch were
meeting with journalists to let them know they were in the game
where the presidential campaign was concerned. And given the
contacts they had built up over the years, they started at the top.
That July, when the Democratic Party gathered in Philadelphia to
nominate Hillary Clinton as its candidate, Simpson and Fritsch
traveled there and met with the executive editor of *The New York
Times*, Dean Baquet, and another senior manager, Matt Purdy,
who oversaw investigative projects. They said they had a stock-
pile of information about Trump's connections to Russia, Paul
Manafort, and other Trump associates that they would be happy
to share with *Times* reporters. Around the same time, they met
in New York with the editor of *The New Yorker* magazine, David
Remnick, as well as an editor at Reuters, the wire service. Simpson
and Fritsch didn't tell those editors anything about Christopher
Steele or the dossier but by then they had started getting memos
from Steele and were trying to plant some of that information in
the news media.

Their first effort involved a little-known Trump campaign advisor named Carter Page. In 2016, Donald Trump appointed Page,
a small-time investment banker and energy consultant, to a hastily
assembled group of foreign policy advisors after he realized that
presidential candidates were supposed to have such panels. And
Page, who was tall, gangly, and wore a trademark bucket hat, soon
drew the attention of journalists because he had previously worked
in Russia and supported the dropping of economic sanctions imposed against Moscow. Just before the Democratic convention,
Steele sent a report dated July 19 to Fusion GPS about Page
that was titled "RUSSIA: SECRET KREMLIN MEETINGS

ATTENDED BY TRUMP ADVISOR, CARTER PAGE IN MOSCOW." He wrote that Page had met during a recent trip to Moscow with Igor Sechin, a top Putin ally and the head of Russia's national oil company, to discuss the dropping of sanctions against Sechin. Steele also described a separate meeting between Page and a Kremlin official, Igor Divyekin, during which Page was told that Moscow had "kompromat" on both Hillary Clinton and Trump and that Trump should bear that "in mind in his dealings" with Russia.

On July 26, 2016, a few days after Steele's report arrived at Fusion GPS, Page got a text message from a reporter with *The Wall Street Journal*, Damian Paletta. In it, Paletta claimed that some of his colleagues at the paper had heard about the consultant's trip to Moscow and that he wanted to run some of those details past Page for comment. "I have always appreciated your candor and I just wanted to be straight with you and ask about some things our reporters have picked up," Paletta wrote. He then posed a series of questions to Page that precisely mirrored the information that Steele had just sent to Fusion GPS.

Paletta asked the consultant to confirm that he had met in Moscow with Igor Sechin and discussed a dropping of U.S. sanctions. He also asked whether the Kremlin official, Igor Divyekin, had told Page that Moscow had "kompromat" on Clinton and Trump.

"Guidance? Thoughts?" Paletta asked.

Page reacted as if dumbfounded by the reporter's questions. "Divyekin?" he wrote. "I just Googled that name and nothing is coming up." He also told Paletta that the kompromat allegations and the claim that he discussed a lifting of sanctions with a Kremlin insider sounded slanderous. "It is such a ridiculous idea that it almost warrants a double-no-comment," he wrote back.

After Steele's reports were posted by BuzzFeed, Page realized that Paletta's questions had been fed to him by former colleagues at Fusion GPS. "I think you guys were pawned that defamatory info by your former @WSJ colleagues at Fusion GPS," Page wrote in a tweet to Paletta. The reporter didn't respond.

Christopher Steele said that all the information in the Trump/ Russia memos he sent to Fusion GPS was gathered through one key informant, or "collector" as he put it, that Orbis Business Intelligence had long used and who had a network of reliable sources inside Russia. Glenn Simpson and Peter Fritsch said Steele never disclosed his collector's name to them but they wrote in their book that he described him as "one of the finest he had ever worked with" and added he was "an individual known to U.S. intelligence and law enforcement." He was "a remarkable person with a remarkable story who deserves a medal for service to the West," Steele had declared.

Steele provided Fusion GPS with a key containing the names or descriptions of the sources from whom his collector was gathering information. "That key identified, sometimes by name, no fewer than seven sources for the report's shocking hotel scene," Simpson and Fritsch wrote in their book. "It also detailed the relationships of numerous other sources to senior people close to the Kremlin and Putin. It was an impressive roster of people in and out [of] government."

IN THE SUMMER OF 2016, not long after Damian Paletta reached out to Carter Page, Glenn Simpson called a colleague of Brian Ross, the on-air correspondent for ABC News, and urged that the network do a piece about a self-styled real estate broker who claimed he was close to the Trump Organization. A few months earlier, the broker, Sergei Millian, had told RIA News, a Russian outlet, that he was the exclusive sales rep for Trump properties in

Florida sold to Russian buyers and described himself as someone who knew about Trump's taste for young Slavic women.

Trump's personal lawyer, Michael Cohen, denied the assertions but Millian quickly drew interest from journalists digging into Trump's Russian ties. Simpson told Ross and an ABC News reporter that the broker had started an obscure trade group, the Russian-American Chamber of Commerce, and that he had changed his name to Sergei Millian from Siarhei Kukuts after he emigrated to the United States from Belarus, a former part of the Soviet Union. Then Simpson dropped the hook. ABC needed to move quickly, Simpson said, because there were suspicions that Millian was a Russian operative. "He said he thought that Millian was a spy and he was going to leave the country," Ross said later.

Like other journalists, Ross didn't know anything in mid-2016 about Christopher Steele or the dossier. He was also unaware that Simpson and Steele were double-teaming Millian. After the RIA News piece, Steele secretly dispatched his collector to approach Millian using the guise that he wanted to discuss a potential real estate deal with him. Meanwhile, Simpson apparently hoped that Ross could get Millian to admit on-air that he was a Russian operative with Trump ties.

When contacted by ABC News, Millian agreed to do an interview. At first, Ross let him carry on about how he had gotten scores of Russians to buy Trump properties and describe the developer's taste in women. Then the reporter got down to business. "Are you involved in any way with Russian intelligence agencies?" he asked. Millian grinned slightly and replied, "Absolutely not." Ross reframed the same question. "You are saying you are not a Russian spy," he asked again, but Millian continued to deny any links to Moscow.

ABC News didn't immediately run that interview but in the

fall of 2016 a number of outlets did stories about Millian. Years later, Steele testified during a lawsuit that the author of a 2016 article in the *Financial Times* about Millian, Catherine Belton, was a "friend" of his, though he insisted he hadn't leaked information to her. But Glenn Simpson was also a friend of Belton's and, soon after her article about Millian appeared, she did a favor for him. In the fall of 2016, Belton met the son of a former executive of Deutsche Bank who was hoping to sell potentially incriminating documents about the financial institution and she told the man, Val Broeksmit, that she knew someone who might be interested. Broeksmit soon got a call from Simpson, whom Belton had described to him as a "really great guy" that was doing oppo research on Trump and Deutsche Bank. Simpson gave Broeksmit about $4,000 and paid for him to fly to the Caribbean, where Simpson and another Fusion GPS operative reviewed his documents about Deutsche Bank, a major lender to Donald Trump.

IN THE FALL OF 2016, with the presidential election just a few months away, select journalists arrived at the Tabard Inn, a small hotel in Washington, D.C. The hotel was a twenty-minute walk from the White House and offered a very different kind of vibe from other hotels in the nation's capital. Its rooms are small and filled with antiquated, worn furniture and it is also probably the last hotel in the area that doesn't have an elevator. Those hardships aside, the Tabard was charming and appealed to people with bohemian tastes. Simpson had chosen it as the setting for face-to-face meetings between journalists and someone they had never heard about before—Christopher Steele.

By then rumors were circulating about a dossier that contained damaging material on Trump. For their part, Steele, Simpson, and Peter Fritsch had been trying for months to get the FBI and other

U.S. government officials to pay more attention to the former MI6 agent's reports. And in July, Steele had grown so concerned by the information he was receiving about ties between the Trump campaign and Russia that he contacted Mike Gaeta, the FBI agent in Europe with whom he was working as a bureau source, and began giving him the memos he was sending to Fusion GPS.

When the FBI didn't appear to be interested, Simpson and Fritsch decided to step up their media offensive. The meetings at the Tabard Inn took place during trips that Steele made to Washington, D.C., in September and October of 2016. Simpson chose the journalists who attended. Most were his friends, former colleagues, or people to whom he had shopped stories in the past. They tended to be journalists who specialized in reporting about politics or national security rather than Russia and, as a result, were far more familiar with the machinations of Washington than Moscow.

The group included Michael Isikoff, Simpson's old friend who was now working for *Yahoo! News*; Jane Mayer, the *New Yorker* writer who had worked at *The Wall Street Journal* while Simpson was there; and Matthew Mosk, a Washington-based correspondent for ABC News. Two reporters with *The New York Times*, Eric Lichtblau and David Sanger, also met with Steele at the Tabard, as did reporters with CNN. While in the nation's capital, Steele went to the offices of *The Washington Post* to speak with a group of reporters there but was unhappy about being seen in one of the paper's conference rooms. "Don't you have any meeting space without glass walls," he inquired, according to a later article in the *Post*.

Steele's sessions at the Tabard Inn had the flavor of a publicity campaign for a new movie where the film's stars endure a series of meetings with the media. Each news organization, to avoid bumping into a competitor, was given a specific time to arrive at the hotel. A spread of food and beverages awaited inside a meeting room. Under ground rules set by Simpson, the meetings were held "off the record," meaning that journalists, absent subsequent approval, couldn't disclose that they had met with Steele or publish anything he told them.

Simpson served as master of ceremonies and started each meeting the same way. He introduced Steele and described his career as a former British intelligence agent who had served in Moscow and elsewhere, adding that as a private operative he worked hand in glove with the FBI to break open the FIFA soccer bribery case.

Then Steele took over. When Simpson talked to reporters about Trump's Russian entanglements, he sounded like someone trapped in a fever dream. Steele, by contrast, was calm, understated, and professional. He told reporters that his reports contained preliminary information—an assemblage of rumors, hearsay, and gossip that spies and operatives refer to as "raw intelligence"—that needed to be confirmed in order to determine whether it was true or not. But he was adamant that everything he had uncovered showed that Trump, his staffers, and the Kremlin were engaged in a hydra-headed conspiracy to rig the 2016 presidential election. "He described Trump as a kind of Manchurian candidate," recalled one reporter who attended those sessions.

SOME OF THE JOURNALISTS who met with Christopher Steele at the Tabard Inn thought he was right. Since the start of the 2016 campaign, Trump had kowtowed toward Vladimir Putin with a subservience that suggested that the Russian leader had a hold over him. There also seemed to be plenty of signs pointing to collusion between the Trump campaign and Moscow. By the fall of 2016, for instance, WikiLeaks was unleashing a nonstop barrage of hacked Clinton campaign emails and Russian operatives were deploying social media bots aimed at swaying American voters toward Trump.

However, believing that the Trump campaign was colluding with the Kremlin was one thing: proving it was very different. And for the reporters who met with Steele at the Tabard Inn, deciding what to do next was a big problem. Even if a journalist believed Steele, there was no apparent way to independently confirm what he had said. They didn't know the identity of his sources, though some journalists assumed they were contacts he had developed while at MI6. That would prove very wrong.

IN THE AFTERMATH OF the Tabard Inn meetings, only one article was written by a journalist who attended those sessions. It was by Michael Isikoff of *Yahoo! News* and was about Carter Page, the energy consultant and Trump foreign policy advisor. The piece bore the headline "US Intel officials probe ties between Trump advisor and Kremlin" and was accompanied by a photograph of Page.

"U.S. intelligence officials are seeking to determine whether an American businessman identified by Donald Trump as one of his foreign policy advisers has opened up private communications with senior Russian officials—including talks about the possible lifting of economic sanctions if the Republican nominee becomes president," Isikoff reported.

BESIDES MEETING WITH JOURNALISTS in Washington, D.C., Christopher Steele went to the State Department to urge officials there to act on his reports. The meeting was arranged by Jonathan Winer, the lawyer once targeted by Kroll operatives at the behest of the con man, Allen Stanford. Winer, who was a money-laundering expert, was a Washington insider who revolved in and out of government depending on which party was in power. A

Democrat, he had been at a law firm during the Bush administration that did lobbying for clients that included Bill Browder, the investor, and Oleg Deripaska, the oligarch. He and Steele became acquainted during that period and reportedly passed on business leads to each other. Then, in 2013, when Winer joined the State Department, he became Steele's contact inside it. Over the next three years, he distributed more than one hundred reports that Orbis Business Intelligence produced for private clients to State Department colleagues. Winer, who had worked as a CIA consultant, was close to Glenn Simpson and journalists in Washington, D.C., and, when reporters looking into the dossier called him, he vouched for Steele.

In the fall of 2016, Winer arranged for Steele to meet with a State Department official named Kathleen Kavalec. During their discussion, Steele described to her the scope of the investigation he had undertaken on behalf of Fusion GPS and how the firm was gunning to get his findings out before Election Day. He told Kavalec that the Kremlin was colluding with Trump campaign officials and that the Kremlin had "kompromat" on Trump in the form of the pee tape. "Steele said he is persuaded the story about prostitutes was accurate because they had their source speak with hotel contacts who confirmed," Kavalec wrote in notes she made of the meeting.

Steele also told her about Carter Page and Sergei Millian and brought up another subject on which he was working for Fusion GPS. His sources had told him, he alerted Kavalec, that a computer server owned by Alfa Bank—the same Russian bank on whose behalf Diligence had pulled off its James Bond–style caper a decade earlier in the Bahamas—appeared to be secretly exchanging messages with a computer server in the United States affiliated with the Trump Organization. Kavalec recorded Steele's comments about the servers in her notes. She had no idea that Glenn Simpson and Democratic Party lawyers were then furiously marketing the same story to journalists in the hope of dropping it on Trump as an "October surprise."

AROUND THE TIME OF his State Department meeting, Steele also took time to pursue the interests of Oleg Deripaska, the Russian oligarch. Early in 2016, the former MI6 agent sent an email to Bruce Ohr, the Justice Department official, insisting that Deripaska wasn't a Kremlin "leadership tool," a view that sharply conflicted with the government's assessment of the oligarch. Several months later, he reached out again to Ohr, this time about one of Deripaska's businesses. By 2016, Ukraine's pro-Putin president, on whose behalf Paul Manafort had worked, had been ousted and the country's new leaders were threatening to nationalize an aluminum smelter that Deripaska's company owned there. The oligarch's London lawyer had asked him, Steele told Ohr, to alert U.S. officials to the development in the hope that they would not support the Ukrainian move. Deripaska's lawyer believes "that the Ukrainians have politicised what is a commercial dispute and may try to suck in other governments sympathetic to them," Steele wrote. "Naturally, he wants to protect the client's interests and reputation."

AS ELECTION DAY APPROACHED, everyone, including Donald Trump, believed that Hillary Clinton would be the next president. Then, with a little more than a week to go, the head of the FBI, James Comey, intervened. He sent a letter to Congress announcing that the agency was reopening its investigation into whether Clinton had mishandled government secrets while serving as secretary of state by using a personal computer server for official business.

Several months earlier, Comey had closed that inquiry and, while criticizing Clinton for her actions, he hadn't accused her of any wrongdoing. Trump had been using the inquiry as a way to hammer Clinton and now Comey, at the election's eleventh hour, had given that cudgel back to him.

Simpson and Steele became increasingly frantic to make the dossier's information public. One journalist recalled getting a late-night phone call from Simpson who sounded agitated and possibly drunk or stoned. He offered to fly up to New York and give a copy of the dossier to the reporter but wanted a promise first. "If I show it to you, I need a commitment that you'll do a piece on it," he said.

Simpson opted for another tactic and arranged for the Washington bureau chief of *Mother Jones* magazine, David Corn, to interview Christopher Steele. Simpson hadn't invited Corn to the meetings at the Tabard but now he had him come to the offices of Fusion GPS to review Steele's memos and interview the ex-MI6 agent over Skype. Bruce Ohr, the Justice Department official, wrote in a memo that Simpson told him he viewed Corn's interview as a "Hail Mary pass" to generate publicity for the dossier.

Corn still couldn't identify Steele by name in his article. But it was agreed that he could treat whatever Steele told him as "background," meaning he could attribute his comments to an unnamed "former Western intelligence officer" or a "former spy." Corn's article appeared a few days after Comey's decision to reopen the Clinton investigation and it carried the type of headline for which Simpson and Steele had hungered: "A Veteran Spy Has Given the FBI Information Alleging a Russian Operation to Cultivate Donald Trump."

Corn reported that the "former spy"—Steele—had told him "there was an established exchange of information between the Trump campaign and the Kremlin of mutual benefit." The article cited verbatim passages from Steele's memos to Fusion GPS, though Corn didn't identify Simpson's firm by name. Instead he described it as a corporate intelligence company involved in political opposition research.

Corn cited the money line from Steele's first report to Fusion GPS about how Moscow had been "cultivating, supporting and assisting TRUMP for at least 5 years." Steele, in his guise as a "former spy," was also quoted in the article as saying that the FBI was investigating his reports. "It's quite clear there was or is a pretty substantial inquiry going on," he said.

When *Mother Jones* posted the piece, FBI officials quickly realized that Steele was the "former spy" in question. They contacted Steele and terminated the bureau's relationship with him, saying he had violated a central tenet of their agreement—namely, that he wouldn't talk to the media about FBI investigations. Bureau officials weren't aware that he had been doing so for months but his fingerprints on the *Mother Jones* piece were so obvious they couldn't miss them.

The FBI agent, Mike Gaeta, who had served as Steele's handler since 2013, testified that he was shocked by Steele's behavior. Gaeta said that he had long thought of the former spy as professional and apolitical and that his only bias was his anti-Russian feelings. But his opinion changed, Gaeta said, when he asked Steele why he had talked to *Mother Jones*. The former operative replied that he had been angered by Comey's decision to reopen the FBI's investigation of Hillary Clinton. "I was completely taken aback because it was so unprofessional," Gaeta remarked.

For his part, Steele would insist that he hadn't agreed while working with the FBI not to speak with the media and pointed out that the FBI had struck its deal with his firm, Orbis Business Intelligence, rather than with him personally. It was an explanation that made him sound less like a hero driven by conviction than a bureaucrat.

IN THE WANING WEEKS of the 2016 campaign, Glenn Simpson and other operatives pushed reporters to do a story about Alfa Bank. After Russian hackers broke into Clinton campaign servers, suspicions quickly emerged that election-related cyberattacks by Kremlin-linked operatives were widespread. Computer experts examined databases of computer traffic and stumbled onto something strange. A computer server owned by Alfa Bank was constantly "pinging," or electronically trying to connect with, a server associated with the Trump Organization. It wasn't clear what was going on but the volume of apparent reach-outs between the computers was huge and computer experts alerted an attorney at Perkins Coie, the Clinton campaign's law firm, who specialized in cybersecurity.

In court testimony, Christopher Steele said that Glenn Simpson and the Perkins Coie lawyer, Michael Sussman, told him during an August 2016 meeting about the suspected "pinging" between the Alfa Bank and Trump servers. Then Simpson specifically directed him to gather information about any connections between three oligarchs who controlled Alfa Bank and the Kremlin, Steele testified.

It was soon after Steele's August meeting with Simpson and Sussman that journalists started hearing about the Alfa Bank "pinging" story. Eric Lichtblau, one of *The New York Times* reporters who met the ex-MI6 agent at the Tabard Inn, was approached in September by a lawyer offering to connect him with an expert who knew about the "pinging." Lichtblau, a national security reporter, never disclosed the lawyer's name. Long afterward, Lichtblau would say that he didn't think Glenn Simpson knew about the "pinging" story at the time he was first approached about it. But given the timing of Simpson's meeting with Steele, it was likely that the Fusion GPS operative was standing offstage, pulling strings.

WHEN ERIC LICHTBLAU STARTED pursuing the Alfa Bank story, he soon came to believe that the servers were secretly communicating. Other staffers at the *Times* were more skeptical. They included another reporter, Steven Lee Myers, a onetime Moscow correspondent who was assigned to work with Lichtblau, as well as the newspaper's editor, Dean Baquet, who didn't quite know what to make of what the servers were supposedly doing.

Meanwhile, in anticipation of the *Times* story, Clinton campaign officials reportedly prepared a video about the Trump/Alfa Bank connection to launch on social media when it was published. But as October started to draw to a close, the newspaper still hadn't produced an article and Simpson apparently began bringing it elsewhere. One online news organization, *The Intercept*, reported that it was among a handful of organizations that also included *The Washington Post* and *Vice News* that were approached with the Alfa Bank story but decided to take a pass.

Eventually, a reporter at *Slate* bit. That journalist, Franklin Foer, wrote an article about the Alfa Bank–related issue that was published about a week before the 2016 presidential election. Its headline read, "Was a Trump Server Communicating with Russia?" Days later, Foer was forced to write a follow-up piece reflecting all the blowback he had gotten, including comments from computer experts who argued that the "pinging" between the two servers wasn't anything nefarious but digital white noise. And when the *Times'* long-awaited "pinging" article was published, it wasn't one that Simpson or Hillary Clinton's campaign was expecting. The headline on it read "Investigating Donald Trump, FBI Sees No Clear Link to Russia."

That headline—much derided afterward—failed to accurately distill the article's contents. Way down in the article, it reported that federal agents were examining possible collusion between the Trump campaign and Moscow and that Paul Manafort was among the subjects of their probe. However, FBI officials were quoted in the article as saying they felt fairly confident that the "pinging" between servers at the Trump Organization and Alfa Bank reflected a glitch or a random exchange of marketing spam, the position that the companies took.

Simpson flipped out. He called several reporters at the *Times* and told them he considered the piece an "abomination." Years later, Steven Lee Myers would remember how Simpson had said that he and his *Times* colleagues had blown the story. And Myers would recall something else Simpson said to him. However the election turned out, Simpson remarked, it wasn't going to affect his bottom line. "He said, 'Either way, I get paid,'" Myers recalled.

MANY JOURNALISTS WOULD BELIEVE that Donald Trump defeated Hillary Clinton thanks to help from Moscow. It was a tempting conclusion to draw. Trump lost the popular vote and his margins of victory in traditional Democratic states such as Pennsylvania and Michigan were so tiny that it was appealing to think the difference was the result of foreign interference, rather than Hillary Clinton's shortcomings as a candidate or blue-collar voter fatigue with Obama administration policies.

There is really no way of knowing how big a role, if any, Moscow's meddling played in Trump's triumph. But one thing would become clear. Glenn Simpson, Peter Fritsch, and Christopher Steele had an "October surprise" in their hands they decided not to drop.

WHILE REPORTING ABOUT ALFA Bank, Eric Lichtblau of *The New York Times* learned that the FBI had opened a counterintelligence investigation into allies of Donald Trump. But he and other journalists weren't aware of its scope and didn't know something that would have made headlines: The FBI had met with Christopher Steele to try and convince him to cooperate with the probe. In October 2016, a team of FBI supervisors and agents met with Steele in Rome and told him about a secret counterintelligence investigation the bureau had launched that summer into potential ties between several Trump staffers and Moscow. The FBI inquiry was dubbed "Crossfire Hurricane."

The public learned about the investigation's existence in March 2017, five months after the presidential election. But it had started nine months earlier when a Trump campaign advisor, George Papadopoulos, made an offhand comment to a foreign diplomat that the Kremlin had hacked into Clinton campaign emails.

When the FBI team met with Steele in Rome, the inquiry had expanded to include an examination of three other Trump associates: Carter Page, Paul Manafort, and Michael Flynn, a retired general who briefly became President Trump's national security advisor. After reading the Trump/Russia memos that Steele had sent to Fusion GPS, FBI officials decided to try to convince him to connect them with his sources so they could assess their credibility.

The FBI offered to pay Steele "significantly" for help on the investigation. But the former MI6 operative didn't seem eager to play ball. During the meeting, Steele sat with his arms folded and his body language suggested that he was "going to be difficult to handle," an FBI official noted. He was also reticent about providing too much information to the bureau about his sources, citing his existing relationship with Fusion GPS and his concerns about the safety of his informants. "Maybe I can go back to the hotel [in Russia] and get the manager for you to meet to talk about the prostitutes being there," he reportedly told the FBI. The only person he described at length during his discussions with FBI agents appeared to be Sergei Millian, the real estate broker whom Steele said he considered to be a "boaster," an "egotist," and someone who might "engage in some embellishment."

Steele and FBI officials came away with very different understandings about what they had agreed to during the meeting, according to the Justice Department's account of it. Those differences became irrelevant. A few weeks after the Rome meeting, James Comey announced the reopening of the FBI's inquiry into Hillary Clinton, setting off the events that led Steele to speak to a *Mother Jones* reporter, and the subsequent decision by the FBI to end its relationship with him.

Simpson and Fritsch knew about Steele's meeting in Rome but neither they nor Steele told journalists about it. After the 2016 election, one top Clinton-campaign official would tell Jane Mayer of *The New Yorker* that if he had known beforehand that the FBI was investigating Donald Trump he would have "been shouting it from the rooftops!" In their book about the dossier, Simpson and Fritsch wrote that they decided not to disclose the FBI's meeting with Steele because they feared their Democratic Party clients might alert journalists about it and, in the process, jeopardize the inquiry.

"Simpson and Fritsch decided that if Hillary's campaign operatives got wind of a possible FBI investigation, it might be unable to resist the temptation to leak it to the press," they wrote.

It's impossible to know, just as it is in the case of Russian bots, how the news of a Justice Department probe of Donald Trump's allies would have affected voters. But the decision by Simpson and Fritsch not to disclose Steele's Rome meeting was remarkable.

Trump wouldn't have hesitated to blast out the news of a similar inquiry into Hillary Clinton. And during a presidential election that was decided by a small margin of votes in a handful of states, their decision to keep Steele's meeting secret may go down as one of the biggest blunders in the history of "oppo."

CHAPTER 10

OUTED, EPISODE 1

OTTAWA, 2016

Around the same time in 2016 that Christopher Steele was meeting with journalists at the Tabard Inn, Rob Moore, the ex-comedy producer turned private spy, was staying at a small, eco-friendly lodge in Ottawa called Angela's Bed & Breakfast. He had been in Canada for a week talking to journalists, politicians, and labor leaders about a documentary he wanted to make that would expose the asbestos industry and was seeking funds to start an anti-asbestos charity.

Four years had now passed since Moore began working for K2 Intelligence as an undercover operative and infiltrated the network of activists headed by Laurie Kazan-Allen. Over that time, he had met scores of public health advocates and attended numerous conferences worldwide. No one suspected he was a private spy and nothing suggested that anyone ever would.

For K2 Intelligence, the asbestos case was generating plenty of profits and perhaps this is what blinded Moore's handler, Matteo Bigazzi, to the game that Moore was running on him. The former comedy producer had decided that he wanted a new role, this time as the kind of activist that the firm was paying him to spy on. Better yet, he saw himself as a whistleblower who would expose not only the best-kept secrets of the asbestos trade but also those of the corporate intelligence industry.

Moore, however, was unwilling to let go of his lucrative career as an operative until he launched a new one. He and a director friend in London were pitching producers there about film projects based on Moore's adventures in the intelligence trade. One would be a straightforward exposé of the asbestos industry. The other would be more of a dramatic production where Moore would star as a private spy coming in from the cold. The film would follow his activities as a K2 Intelligence operative leading up to a public health conference where he would unmask himself as a corporate spy and whistleblower while the camera was rolling. But so far, Moore hadn't found any buyers.

Not long before his trip to Canada, Moore had glimpsed another possible escape route—one that ran right through K2 Intelligence. In late 2015, another executive at the private spying firm had approached him about working on a new case involving an oil industry client. Moore would later say that the executive explained that prosecutors in several countries were investigating whether two energy giants—Shell, a Dutch company, and Eni, an Italian firm—had bribed officials in Nigeria to win drilling rights there.

Moore was told that Global Witness, the anticorruption group, had first uncovered evidence of the suspected payoffs and was now working with prosecutors to develop their cases. The K2 Intelligence executive wanted to know, Moore said, if he was interested in donning his guise as an investigative journalist and infiltrating

Global Witness in order to learn where the investigations were heading.

Soon afterward, Moore was in a London courtroom observing a hearing connected to the Nigerian case. When it ended, he approached a Global Witness official and introduced himself as a journalist and a filmmaker. "I've got a story that I think might be of interest to Global Witness, actually two, and I would love to get your thoughts on who I should be talking to," Moore wrote in a follow-up email.

ROB MOORE KNEW THAT Global Witness had a tradition of working with whistleblowers who had information about suspected corporate and political corruption. His idea was a twist on that theme. He would tell Global Witness that K2 Intelligence wanted him to spy on it and then offer to act as a double agent who would keep the group informed about what the firm's client wanted him to find out.

Moore spent weeks debating the plan's pros and cons with friends. Some told him it was too risky and that he couldn't trust Global Witness to keep his identity secret. But emotionally Moore's life as a private operative was taking a toll. When he had lost his television job years earlier, Moore had suffered a breakdown. Then in 2013, not long after he had started working for K2 Intelligence, he started experiencing blinding headaches and was hospitalized. Tests were run to determine if he had a brain tumor but it turned out that he was suffering from cluster headaches that a doctor diagnosed as related to stress.

The woman with whom Moore was now living had urged him to stop working for K2 Intelligence. Moore had come to believe the firm's asbestos client wasn't an "American investor," as his handler, Matteo Bigazzi, had told him, but a little-known investment

firm based in Singapore called the Kusto Group, which was controlled by a Kazakh oligarch.

ROB MOORE WAS BEGINNING to worry too. In the spring of 2016,
he traveled to Hanoi to speak at a meeting of public health
officials and local journalists. He had convinced Matteo Bigazzi
to pay for the trip because Vietnam was one of the countries in
Southeast Asia considering an asbestos ban and Moore said that
attending the conference would help maintain his cover story as
a journalist.

At the meeting, Moore gave his whistleblower persona a test
run. One of his talks at the Hanoi conference was titled "The Duty
of Journalists to Protect Public Health." He listed those duties in
a PowerPoint slide as the responsibility, among others, "To speak
truth to power" and "To challenge corrupt practices." He also cas-
tigated an industry-funded scientist who, during a recent visit to
Vietnam, had played down the risks of chrysotile asbestos. And
he showed his audience a slide containing photographs of the
chairman of the Kusto Group, Yerkin Tatishev, and some of his
associates.

Two days later, Moore's cellphone rang and the name "ZAD"
flashed in the caller ID. It was his code name for Matteo Bigazzi.
"What the fuck is going on!" the K2 Intelligence executive
shouted. "We've heard that there has been a vicious attack on the
industry."

Moore was jolted back to reality. He said he improvised and
told Bigazzi that an American asbestos expert who had planned
to speak about the Kusto Group at the Hanoi meeting had been
forced to cancel his trip and insisted that he give his talk for him.
Bigazzi seemed mollified by the explanation but offered Moore
some advice. It was fine for him to pretend he was an activist but

his act had limits. "I don't want to hear you are speaking about the industry's corruption because then you are putting your head above the parapet," Moore said Bigazzi told him.

That evening, Moore heard a knock at the door of the Airbnb apartment in Hanoi where he was staying. He had given its address to his companion but no one else, not even Bigazzi. When Moore peered out through the door's peephole, the hallway looked empty. When he asked who was there, no one responded. He panicked and suspected that a goon dispatched by the asbestos industry was lurking outside, waiting to beat him or worse. The apartment was on the building's third floor but Moore could see from the bathroom window that the rooftop of the neighboring building was not far below. He imagined himself jumping onto it and then shimmying down to the street, where he would steal a moped and make a Jason Bourne–style escape.

He called the apartment's Airbnb host but no one picked up so he left a message on the answering machine. For the next two hours, he frantically deleted documents on his laptop related to Kusto or sent them to encrypted email accounts he had created. When the Airbnb host finally returned his call, Moore's Jason Bourne moment became more like one from Mr. Bean. His visitor, he was told, had been the apartment's maid. He hadn't seen her through the door's keyhole because she was too short to reach it and she had been too shy to respond to him when he had called out.

NOT LONG AFTER HIS return from Vietnam, Rob Moore walked into London's St. Pancras train station, a mammoth Victorian-style edifice. A founder of Global Witness, Simon Taylor, sat at a restaurant table waiting for him. Taylor was a solidly built man in his early sixties who had overseen many of Global Witness's biggest cases. When accepting an award given to the group, he likened his experiences to events in the 2005 film *Syriana*, a thriller that starred George Clooney as a CIA agent in the Middle East who is drawn into a web of corruption. "For me, watching this film was a surreal experience because I have met and investigated many of the people like those in the film," Taylor said.

After introductions, Moore told Taylor about his role as a contractor for K2 Intelligence and how he had been spying for several years on the anti-asbestos advocates. Moore said he knew they would be shocked when they learned his secret but emphasized that he planned to unmask himself as soon as his planned documentary about the asbestos industry was done. When that happened, he was certain that Laurie Kazan-Allen and her colleagues would understand that he had been on their side from the start.

"I've worked with these people for three or four years," Moore explained to Taylor. "They care about me and trust me and it's been the hardest thing, to know what I've done. The only way I've been able to do it is knowing that this was where it would be going."

He then told Taylor about his new assignment to infiltrate Global Witness. He said he didn't know the identity of K2 Intelligence's client though he has his suspicions. Then he made his pitch to become a double agent. "I'll give you as much as you want," he explained. "I'll be able to share everything with you."

Moore reached into his bag and pulled out a book that he gave to Taylor. It was titled *Agent Zigzag* and told the real-life story of a small-time English crook named Eddie Chapman, who was trained by the Nazis as a spy during World War II but then flipped and became a double agent for British intelligence.

"It's a stunning story," Moore went on. "It's basically what I'm pitching to you now."

If Global Witness was interested, he told Taylor, there were

things he needed from the group. Most important, it had to protect his identity because he feared that K2 Intelligence might try
to destroy him and his family if it discovered his betrayal. Also,
to convince the firm and its client that he had penetrated Global
Witness, Taylor's colleagues would need to feed him tidbits about
the bribery investigations that he could report back. "I'm very
happy to stay in this," Moore said. "I feel like I've found a purpose."

Taylor said he was intrigued with Moore's proposal but needed
to discuss it with his colleagues. However, he assured Moore that
the last thing he wanted was for him to be exposed. "I'm not interested in that at all," Taylor explained. "I'd rather not do anything at
all, if that's the possible outcome."

ROB MOORE HEARD BACK from Simon Taylor not long before
he departed for Canada. Taylor wrote that the leaders of Global
Witness had spent weeks debating whether to go forward with
Moore's proposal but ultimately decided that its risks outweighed
any potential benefits. They were concerned, among other things,
that Moore might turn out to be a triple agent who would keep
K2 Intelligence abreast of what Global Witness was doing while
feeding the anticorruption group useless chum.

There was another, more pressing problem, Taylor wrote. The
anti-asbestos advocates were natural allies of Global Witness and
keeping Moore's undercover role secret would put the group in
the untenable situation of being a party to that deception. "It is
our strong view that the preferred way forward would be that as a
matter of priority, you would take the necessary steps to inform all
those you have deceived of your role," Taylor wrote. "We suggest
that there is no time like the present to do so."

To Moore, Taylor's email read like a carefully crafted "cover-
your-ass" memo that Global Witness could pull out in the event its

interactions with Moore came to light. He didn't take it seriously. Also, he was the person who would set the time and circumstances of his outing and that wasn't going to happen until he got a commitment to make the asbestos documentary he was pitching.

He wrote back to Taylor, saying he expected Global Witness to make good on its promise to protect him until his planned asbestos documentary was done. "For the sake of my family's safety as well as my own and the success of the journalistic endeavors that are now underway, it is essential that Global Witness continue to guarantee the assurances you have given me," he wrote.

WHILE STILL AT ANGELA's Bed & Breakfast, Rob Moore decided to call an activist in London with whom he had been working to set up the anti-asbestos charity. But when the man picked up the phone and heard Moore's voice, he couldn't get off fast enough. "I can't really talk about it now, Rob," Moore later recalled him saying. "I've been told you work for K2 or an agency."

Moore realized immediately that Global Witness had given away the game. He admitted that K2 Intelligence had been paying him but insisted that his goal was to expose the asbestos industry. "I can explain. . . . Give me a chance," he said. The activist told him he was hanging up. "I can't talk to you now," he said. "It is not looking good, Rob."

Moore notified his filmmaker friend in London that his secret was out. The director started running Google searches about Global Witness and told Moore that the anticorruption organization was represented by a large plaintiff's law firm in London called Leigh Day. Moore nearly dropped his phone. He had recently invited two Leigh Day lawyers to become directors of his planned anti-asbestos charity.

At that point, most people would have gone to bed and pulled the sheets over their head. Moore remained convinced he could sort things out. As part of his trip to North America, he had planned to travel to Washington, D.C., to meet with public health activists as well as with a reporter affiliated with the International Consortium of Investigative Journalists. The group was an alliance of reporters that had produced a landmark report in 2010 about the global trade in asbestos.

But news of his ties to K2 Intelligence was spreading quickly among the tight-knit community of anti-asbestos activists. He was supposed to stay at the Washington-area home of an occupational health expert but when he called the man he was told he was no longer welcome there. He flew to the nation's capital anyway to meet with the reporter, who also had been tipped off to Moore's spy business connections. The journalist agreed to take whatever information Moore wanted to share about the asbestos industry in Southeast Asia but had a hard time making sense of it.

WHEN ROB MOORE RETURNED to Canada to get his flight back to London, he received an email from an attorney at Leigh Day, notifying him that Laurie Kazan-Allen and several fellow advocates were planning to file a lawsuit against him. They would seek financial damages for their personal data, which they said was stolen after he had penetrated their coalition. Moore felt physically ill. Not long before boarding his flight, he got a call from Matteo Bigazzi, who was eager to hear about what he had learned during his time abroad. Moore didn't mention anything to him about Global Witness or the threatened lawsuit but told Bigazzi that he was having a migraine headache and would send him a report soon.

When Moore's plane landed in London he was a nervous wreck. A cabdriver he knew picked him up at Heathrow Airport and Moore asked him to make a stop before taking him to his

home. "Can you drive me to my friend's?" Moore asked. "I know
he has some Valium."

ROB MOORE'S LIFE QUICKLY devolved into a legal nightmare. Since
coming back to England, he had tried everything possible to head
off the threatened lawsuit. He had offered to give lawyers at Leigh
Day, the firm representing the activists, all his correspondence with
Matteo Bigazzi, including the messages they exchanged through
"benthiczone.com," the account created to "folder" their emails. A
British lawyer known for winning reprieves for death-row inmates
also went to bat for Moore and tried to get Leigh Day to stand
down.

The law firm wasn't interested and Moore soon walked into
the Royal Courts of Justice, a sprawling building in London. He
made his way down a massive arched hallway until he reached
a café, where he met two lawyers. That morning he had decided
to forgo taking Valium because he wanted to be alert in case the
judge spoke to him. Still, as the hearing began, he felt numb and
the scene in the courtroom unfolding in front of him seemed surreal, as though it was taking place underwater. Lawyers were also
speaking in a jargon he barely understood.

At the hearing's end, attorneys on both sides of the case agreed
to keep Moore's identity anonymous and he was referred to in
court filings under the initials "DNT." Moore felt a sense of relief, though that quickly ended when lawyers for the activists announced that they planned to add K2 Intelligence and Matteo
Bigazzi as defendants in the lawsuit.

Not long after the hearing, Moore's cellphone rang and the
name "ZAD" appeared. Moore figured that there was only one
reason why the K2 Intelligence executive was calling. He probably
had just gotten served with court papers and was freaking out.

Before flying back to London, Moore had sent the firm a bill for his trip to the United States and Canada, totaling about $10,000. Moore checked his bank account and saw that the payment for his invoice had cleared. He let Bigazzi's call go to voice mail. He had nothing more to say to him.

ROB MOORE'S ANONYMITY DIDN'T last long. An Australian news site published a series of articles about the asbestos incident that named him. "The global spying operation, codenamed Project Spring, was the brainchild of K2 Intelligence and run from its London office," one article by the publication *New Matilda* stated. "It involved placing a corporate spy at the heart of the anti-asbestos movement, which for more than a decade had been building momentum for a world-wide ban on the deadly mineral."

Moore was pilloried in the *New Matilda* pieces as a duplicitous undercover agent and his sister, who was a top executive at the BBC, got dragged into the mess. "Brother of BBC boss 'was paid to spy on asbestos campaign,'" the headline of one British newspaper article read. Moore was soon spending large amounts of money on lawyers and his attorney called him one day to deliver what he believed was good news. Moore said that the lawyer told him that K2 Intelligence had made an offer—it would cover his legal costs if he agreed to align his defense with its position.

K2 Intelligence took the stance that its actions in the asbestos case had been legitimate because its client, whom the firm didn't identify, had reason to suspect that there was a "corrupt association" between the anti-asbestos activists and American plaintiffs' lawyers. But Moore decided he couldn't take the offer because he thought the firm was wrong and believed a trial would provide him with a stage to explain his actions and achieve redemption.

The lawsuit wasn't a great advertisement for the ethics of K2

Intelligence or how it vetted personnel. One of the episode's first apparent casualties was when the firm's motto about how it always did "the right thing" vanished from its website. The asbestos case got some media coverage in England but received little in the United States, and lawyers suing the firm in London limited their action to the firm's operation there rather than filing a similar lawsuit in New York. The decision meant that Jules Kroll and his son Jeremy were kept out of the legal line of fire and wouldn't have to testify about how the firm's top executives monitored intelligence operations.

The case dragged on for a while. Then it was resolved in 2018 through an out-of-court settlement under which K2 Intelligence paid money but didn't acknowledge any wrongdoing. Laurie Kazan-Allen, in a posting on her blog, described the financial terms of the undisclosed settlement as generous. Still, she expressed disappointment that the case had concluded before it had disclosed the truth. (Kusto, the firm headed by Yerkin Tatishev, said that neither it nor its chairman was the client and that the company didn't have any interests in asbestos.) "Unfortunately, the identify of K2's ultimate client . . . remains hidden," she wrote.

Rob Moore repeatedly insisted that he would never agree to a settlement unless it provided him with an opportunity to clear his name. But that never happened and he would later say that he had to accept the settlement's terms or face financial ruin. Afterward, he continued to lash out at Global Witness, Leigh Day, and other people who, he said, had betrayed him. Years later, he was working part-time at night cleaning offices and talking to journalists in the hope that his story as a whistleblower would one day come to light.

WITHIN K2 INTELLIGENCE, THERE was fallout from the asbestos case. While the lawsuit was under way, the firm and Matteo Bigazzi had remained wedded as codefendants. When the case concluded, he left. No reason was given. Still, it was hard to imagine that Bigazzi's superiors were pleased with how he had handled Rob Moore or how the former comedy producer had handled him.

Charlie Carr, the freewheeling K2 Intelligence operative in London, was a different story. He left the firm right around the time the asbestos debacle came to light. His departure marked the end of his three-decade-long association with Jules Kroll during which Kroll chose to overlook Carr's adventures and his dismissive attitude toward his son, Jeremy. Carr was the money machine who had breathed life into K2 Intelligence during its infancy. Now he was starting a new firm with his brother-in-law.

"Charlie Carr has left K2 and is now working for the [Kazakh] Trio," Mark Hollingsworth wrote another private operative.

AROUND THE SAME TIME as Robert Moore's outing, Mark Hollingsworth's double life almost came unraveled when a newspaper, *The Irish Times*, published an article in 2016 disclosing that he had misrepresented himself as a journalist. Hollingsworth had interviewed people for what he said was an article for the London *Sunday Times* but when reporters for the Irish paper contacted editors there they said they hadn't assigned Hollingsworth to do the story. *The Irish Times* reported that Hollingsworth had actually been gathering information for Alaco, a corporate investigations firm in London that used him as a contractor. Executives at Alaco declined to comment but an executive there said the firm often used what she described as "tame" journalists for assignments.

Journalists and activists were also growing suspicious of Hollingsworth's apparent ties to private spying firms working for people connected to a major financial scandal unfolding in Malaysia. By 2016, authorities in the United States, Switzerland, and elsewhere were investigating the role of Malaysia's prime minister, his relatives, and associates in plundering billions of dollars from a government-run investment fund known as "1MDB."

Several journalists had been investigating the 1MDB scandal, including Clare Rewcastle Brown, a former staffer at the BBC who now ran a blog that covered environmental and corruption issues in Malaysia, and reporters at *The Wall Street Journal*. Rewcastle Brown said she received an email from an executive at K2 Intelligence's office in London seeking her help on behalf of one of the firm's new clients who was described as being concerned about the "potential pitfalls" of doing business in Malaysia.

In his note, that operative, Jason Lewis, noted that he was a former journalist. "Any help would be treated in total confidence (with the same degree of trust I adopted throughout my journalistic career) and as deep background," Lewis wrote. "But it would be most useful as I have little idea of the political ins and outs of the political situation in the country."

Lewis also told Rewcastle Brown that K2 Intelligence would be happy to offer her lucrative freelance work writing up reports about Malaysia. She decided not to follow up on his offer, and that was probably a good thing. According to a U.S. government filing, a suspected mastermind of the 1MDB scandal, Jho Low, sought to set up meetings in 2014 with K2 Intelligence.

Soon after getting approached by Lewis, Rewcastle Brown also heard from Mark Hollingsworth. He said that one of his longtime contacts, a man named Nicolas Giannakopoulos who was associated with the University of Geneva, was organizing a conference there to discuss the 1MDB scandal.

Rewcastle Brown wrote on her blog that when she arrived at the Geneva conference something didn't feel right. An email about the meeting said that students and Swiss government officials would take part in the event, but she didn't see members of either group there. Nicolas Giannakopoulos and other organizers of the meeting insisted it was a legitimate conference. Separately, she also soon learned that Giannakopoulos didn't just have a connection to the Swiss university. He was also a hired operative. And Giannakopoulos and Hollingsworth worked with K2 Intelligence on a case in 2015 when the investigative firm was hired by an executive of a major investment fund, Softbank, to dig up dirt on his rivals for a top position there, according to a later report in *The Wall Street Journal*.

MARK HOLLINGSWORTH ENCOUNTERED SOME turbulence as a result of *The Irish Times* article but his path as a journalist/private spy continued uninterrupted, a tribute to his entrepreneurship, the demand by corporate intelligence firms for his services, and a lack of standards among British media organizations. One newspaper where Hollingsworth freelanced, *The Guardian*, adopted a policy in 2011 requiring freelancers to voluntarily divulge potential conflicts of interest. But Hollingsworth insisted that he was never asked by anyone at *The Guardian* or at other media outlets to make those kinds of disclosures.

Hollingsworth soon set his sights on a new cache of sensitive corporate documents that he would shop to private operatives. The records, known as the "Panama Papers," represented a vast inventory of previously undisclosed corporate filings showing the owners of thousands of shell companies. The documents had been stolen from a law firm in Panama that specialized in setting up shell companies for oligarchs, politicians, and the wealthy. And in time, they made their way to the International Consortium of Investigative Journalists, the same group that issued the 2010 report about the global asbestos trade, and the organization compiled the records into a database. The group was aware that private spies wanted access to the records and put safeguards into place to prevent that from happening. Only media organizations that were collaborating with the consortium had access to the database and, even then, just a few staffers at each outlet were authorized to query it.

Hollingsworth needed to figure out a workaround to get into the Panama Papers and he did it by adopting a strategy he employed through his career as an operative—he put on his journalist's hat. He contacted reporters with whom he had previously collaborated on articles about ENRC, the mining company owned by the Kazakh Trio, and suggested they work together again on a piece about the firm, which remained under investigation by the Serious Fraud Office.

One reporter he contacted was a journalist at *The Guardian*, Simon Goodley, with whom he had written a 2013 article about ENRC. "You may recall that we did stories together on ENRC and so I thought we would revisit this one if you have the time," he wrote. "One of the priorities is to obtain documents from the Panama Papers about the Trio."

Goodley recalled that article. But he didn't know that Hollingsworth, back in 2013, had sent a prepublication draft of the piece to Glenn Simpson in case the Fusion GPS operative wanted to make editing suggestions. The move was the type of breach of professional ethics for which Simpson could have gotten fired during his days as a reporter.

"Below is a very rough draft of our story," Hollingsworth had written Simpson. "Please check for accuracy but also feel free to insert details and material that we have missed but please email me back before the end of today (Thurs) because events are moving rapidly."

In 2016, Goodley told Hollingsworth he wasn't interested in collaborating so the journalist/private spy reached out next to a former reporter at *The Guardian*, Simon Bowers, who now

worked for the International Consortium of Investigative Journalists. Hollingsworth sent him the draft of a proposed article about
the ongoing probe into ENRC, containing blanks where Bowers
could insert information from the Panama Papers. "Clearly, the
story is dependent on PP searches in co-operation with the ICIJ,"
Hollingsworth wrote. Bowers, who was aware of the *Irish Times*
piece about Hollingsworth, ignored the approach.

But the journalist/private spy found other reporters with access to the database who were willing to help. And soon enough,
Hollingsworth was shopping "Panama Papers" records to other
operatives. In one case, he told a private investigator in Washington, D.C., that he needed money up front to get the Panama
Papers documents he wanted. "My source will not accept anything
less than $2,000 for the documents and so please talk to your cli-
ent," Hollingsworth wrote the operative, Richard Hynes. "I think
that is quite reasonable."

Hollingsworth offered far more liberal terms to his old friend
Alex Yearsley, the former Global Witness investigator who had
introduced Glenn Simpson and Christopher Steele. "Please email
me your hit-list of individuals and companies and I will run
searches for you on the Panama Papers database—happy to do
some gratis but I would hope that we can get paid for some of the
docs," he wrote Yearsley.

Another one of Hollingsworth's clients for Panama Papers
documents was a hired spy who was about to promote his own
story to journalists—Christopher Steele. In April 2016, right
around the time Steele started working for Fusion GPS, Hollingsworth told him he was getting a BBC producer to run Panama
Papers searches for companies in which Steele was interested.

Hollingsworth told Steele that he had asked his BBC contact
to search the Panama Papers for an obscure company called Novi-

rex Sales. The shell company's principals, as it happened, included Paul Manafort, the lobbyist whom Steele was then pursuing on behalf of the oligarch Oleg Deripaska.

In that email, Hollingsworth said he hoped his efforts on behalf of the former spy would cancel a debt he owed him.

"Since we spoke on the phone tonight, there is a possibility of more access to the Panama database and so I may get more hits on the second list you sent me," Hollingsworth wrote. "If I have success, then it will resolve my problem of payment on Project Scooter."

CHAPTER 11

OUTED, EPISODE 2

WASHINGTON, D.C., 2016

During the 2016 campaign, Glenn Simpson reshaped the role of private operatives in "oppo" research by hiring a former foreign spy, Christopher Steele, to investigate Donald Trump. After Trump's election, Simpson and Peter Fritsch expanded those boundaries again, this time by transforming "oppo" from a seasonal pursuit into a forever business.

Trump's victory sent Simpson's anxieties into the red zone, and for a time, only a small circle of people knew about the dossier and Fusion GPS's role in it. But Simpson feared that if the news got out, a crazed Trump fanboy might try to attack him and he flirted with the idea of fleeing to Canada.

Instead, in late 2016, with two months remaining before Trump's inauguration as the country's forty-fifth president, Simpson, Peter Fritsch, and Christopher Steele started a new offensive to get the dossier in front of government officials and the media. In mid-November, an associate of Steele's approached Senator John McCain of Arizona and a colleague at a meeting of national security specialists in Halifax, Canada. Sir Andrew Wood, a former British ambassador to Russia, told McCain's colleague, David Kramer, a onetime State Department official and a director of the McCain Institute, about Steele's reports, explaining they detailed collusion between the Trump campaign and the Kremlin and showed that Moscow had compromising information about the new president.

There had long been bad blood between Senator McCain, who was a leading congressional critic of Vladimir Putin, and Trump, who had once dismissed McCain's wartime valor. McCain and Kramer both viewed Trump as incompetent and dangerous and Kramer had voiced his concerns to journalists during the 2016 campaign. Once McCain heard from Wood, he asked his former aide to fly to England to meet Steele.

Steele picked Kramer up at Heathrow Airport, and they drove to the operative's home in Farnham, where Steele gave him an extensive briefing about the dossier and then took him back to Heathrow to catch a return flight to Washington, D.C. Steele didn't give Kramer copies of his memos but showed him a key identifying his collector's informants. The next day Kramer met with Glenn Simpson, who gave him two copies of the dossier to deliver to Senator McCain who was going to pass them on to James Comey, the FBI director.

Meanwhile, Simpson was working another angle. About a week after David Kramer returned from England, Simpson met with Bruce Ohr, the Justice Department official who was close to Steele, at a Washington coffee shop. The men knew each other because Ohr's wife was a fluent Russian speaker and former State Department employee who once worked for Fusion GPS. At the coffee shop, Simpson handed Ohr a computer memory stick containing Steele's memos. He told Ohr that Donald Trump's longtime lawyer, Michael Cohen, had been the "go-between" for the Trump campaign's dealings with the Kremlin.

Right around the time of that coffee shop meeting, David Kramer alerted Simpson and Christopher Steele that James Comey now had the dossier. And soon afterward, Kramer started getting a stream of calls from reporters. Those journalists included David Corn, the *Mother Jones* correspondent who had interviewed Steele, a Washington-based reporter for ABC News, journalists with *The Washington Post*, and reporters for the McClatchy chain of newspapers. They all wanted Kramer to confirm a tip that Comey had the dossier. He was eager to publicize the development, but when ABC News filmed an interview with Kramer he refused to say on camera that it was Senator McCain who had given Comey the dossier, and the interview was never broadcast.

SOME REPORTERS SAID AFTERWARD that it was Glenn Simpson who had alerted them about Comey's receipt of the dossier. The ex-reporter understood the needs of journalists; it was the foundation of his career as a private operative. And with the presidential election over, reporters needed a fresh hook to write about the dossier. A decade earlier, when Simpson was a reporter at *The Wall Street Journal*, his sources had gotten him to "front-run" a report about the investigations that were supposedly under way into Alexander Mirtchev, the Kazakh-connected consultant. Now Simpson was apparently taking a page from that playbook by pressing reporters to "front-run" a piece disclosing that top FBI officials were examining the dossier. One journalist recalled getting a call from Simpson in which the operative insisted that it was time to go with a story about the dossier because a government probe was under way. "He said it was a game changer," that reporter recalled.

NEAR THE END OF 2016, David Kramer arrived at a holiday season party in Washington, D.C., and bumped into an old friend, Alan Cullison, the *Wall Street Journal* reporter who had once been based in Moscow. Kramer had given the dossier to some journalists in the hope they could confirm some of its allegations prior to Trump's inauguration. Kramer told Cullison he had something for him and not long afterward, Cullison was on a flight from Paris to Prague to look into the dossier's allegations about Michael Cohen, Trump's lawyer.

A month before the 2016 election, Steele had sent a memo about Cohen to Fusion GPS that contained one of the dossier's most explosive allegations. Steele wrote that a well-placed source reported that Cohen had secretly met in Prague during the campaign with Kremlin operatives to discuss how to coordinate tactics. The claim, if true, would provide direct evidence of collusion between Trump's inner circle and Vladimir Putin's regime.

"Speaking in confidence to a longtime compatriot friend in mid-October 2016, a Kremlin insider highlighted the importance of Republican presidential candidate Donald TRUMP's lawyer, Michael COHEN, in the on-going secret liaison relationship between the New York tycoon's campaign and the Russian leadership," Steele wrote.

Before leaving for Europe, Cullison had shown Steele's reports to one of his sources, a former official with the National Security Council. The man skimmed through half the memos, before stopping to ask skeptically, "How much of this are we supposed to believe?"

Cullison was skeptical of the dossier's allegations, too. But he, like many reporters, felt they were potentially so serious they had to be chased down. Besides, the idea of a clandestine meeting involving Michael Cohen and Russian operatives didn't seem all that far-fetched. The lawyer was a bully and Mafia-style fixer who once had memorably said that he would be willing "to take a bullet for Mr. Trump."

On Cullison's flight to Prague, an attractive young woman took the seat next to him and soon tried to strike up a conversation. Cullison presumed from her accent that the woman, who had blond hair and looked to be in her late twenties, was Czech. She told him that she was on her way to Prague to visit her boyfriend, whom she described as an older investment banker. Cullison, who was her senior by three decades, replied curtly and tried to stay focused on his work.

When the flight landed in Prague, Cullison said good-bye to his seat mate and headed toward the baggage carousel. There, the woman reappeared and stood next to him while they waited for their luggage. Cullison got his bag, said good-bye again and made his way to an airport coffee shop, where he pulled out a laptop to make a reservation at a Prague hotel. He had traveled in recent days to Amsterdam and Paris to run down other information contained in the dossier and, as a safeguard, would wait until the last minute to make hotel reservations in case someone was monitoring his movements. Suddenly, the woman reappeared and suggested they share a taxicab into town. Cullison agreed and when they arrived at his hotel, she proposed they have a drink at the hotel's bar. Cullison sensed she wanted to move the conversation elsewhere so he steered her out of the hotel and into a cab.

Later, he got a note from the woman saying how much she wanted to see him again. Cullison was still in Prague trying to learn whether Cohen had been there and, after a day of interviews, returned to his hotel room. During his years as a foreign correspondent, he had made a habit of carrying a financial emergency kit with him in case he needed to get out of somewhere fast. It consisted of $2,000 in U.S. bills and two ounces of gold. When Cullison looked into his bag, he saw that the money and the gold were still there. The only thing missing was the dossier.

OTHER JOURNALISTS WHO HAD gotten dossier memos from Simpson tried to confirm its contents, including the story of the "pee tape." ABC News dispatched a Moscow-based stringer to the Ritz-Carlton hotel, where she tracked down the manager of its housekeeping services. The reporter delicately asked the woman if she knew anything about Trump and pee-stained bedsheets. "If anything like that had happened at my hotel, I would have known about it," the housekeeping manager responded.

Another dossier memo claimed that Michael Cohen's father-in-law was a politically connected oligarch in Ukraine who owned a home outside Moscow. But an ABC News stringer reported back that the property in question wasn't owned by Cohen's father-in-law but by someone else who happened to have the same name. Brian Ross, the ABC News correspondent, decided to call Cohen, hoping to get him to confirm the dossier claim that he had gone to Prague to meet Kremlin operatives. "I'm sure I'm not the first one to ask you about this, but have you ever been in Prague?" Ross asked Cohen. The lawyer, seemingly puzzled, replied, "You are the first one to ask me this and I've never been to Prague."

After Ross told him about the allegation, Cohen said that he had been in Capri, Italy, at the time of the supposed Prague meeting. He insisted that apart from having driven once through the Czech Republic he hadn't been in the country for a decade and could prove it because his passport didn't contain an entry stamp. Later, after Buzz-Feed posted the dossier, Cohen offered to come to the studios of ABC News and do an on-air interview and a black car was dispatched to pick him up at his Manhattan apartment. But the lawyer soon called back to say that Trump had nixed the interview and he subsequently was a guest that evening on Sean Hannity's show on Fox News.

Years later, Brian Ross said that he often called Glenn Simpson to tell him when leads from the dossier hadn't panned out. In one instance, Simpson may have taken that information to heart because in one version of the dossier, the material about the supposed Moscow dacha owned by Michael Cohen's father-in-law was blacked out. But more typically, Ross said, Simpson insisted that Steele's information was accurate and offered an explanation why. He said the lack of a Czech stamp in Cohen's passport didn't mean anything because the lawyer could have entered the Czech Republic from Italy or another European Union country, thus avoiding passport control. "Oh yeah, you can go across the border without having your passport checked," Ross recalled Simpson saying.

IN THEIR BOOK ABOUT the dossier, Glenn Simpson and Peter Fritsch laid the blame for BuzzFeed's publication of Christopher Steele's memos at the feet of David Kramer, the McCain aide. Kramer did play a role in the document's disclosure but absent efforts by Glenn Simpson and Christopher Steele to pull the strings of journalists, it's possible the public would have never learned about the dossier or the former MI6 spy.

The chain of events that led to BuzzFeed's publication of the dossier had its origin at a weekend staff retreat held by Fusion GPS in early December 2016 at a mansion north of San Francisco. A reporter for BuzzFeed, Ken Bensinger, who had written about the FIFA soccer scandal, was in the area researching a book about the topic when he got a call from a Fusion GPS staffer inviting him to join the festivities.

Bensinger, because of a vehicle mishap, only arrived at the retreat around 10:30 p.m. It had the ambience of a bros' bacchanal. There were steaks, open bags of potato chips, and plenty of booze. As people headed off to bed, an intoxicated Simpson approached Bensinger to chat. They talked first about the FIFA scandal and Bensinger's planned book. Several months earlier, Bensinger had started tapping Christopher Steele for information about FIFA, and while he didn't tell Simpson that Steele was his source, the Fusion GPS operative was apparently aware of their relationship.

Out of the blue, Simpson began telling Bensinger about the dossier, emphasizing the story of the "pee tape" and describing how Donald Trump had long been a secret asset of the Kremlin. Steele hadn't told Bensinger about his Trump memos and the journalist's appetite was whetted. When Bensinger asked Simpson for a copy of the dossier, he said he didn't have one but Bensinger would later come to believe that Simpson was lying about that, and that his invitation to the Fusion GPS retreat had been a subterfuge to get him wound up about the dossier.

If so, the strategy worked. Shortly after the Fusion GPS retreat, Bensinger called Steele to ask him about the dossier but the former spy acted like he didn't know about it. Bensinger then learned that David Kramer had the dossier and Kramer, after getting contacted by the BuzzFeed reporter, spoke to Steele, who vouched for him. When Bensinger arrived at the McCain Institute on December 29, Kramer allowed him to sit alone in a room to review Steele's reports. Along with taking notes, the reporter used his cellphone camera to photograph each page of the memos and emailed the documents to the offices of BuzzFeed.

By New Year's Day, the news organization had dispatched a team of reporters to try to confirm the material in Steele's reports and Bensinger arrived in London on January 3 to meet with Steele. David Kramer said he met around then with another reporter at Steele's suggestion.

That journalist, Carl Bernstein, had achieved fame decades earlier when he and his then–*Washington Post* reporting partner, Bob Woodward, broke the Watergate story. He was now working for CNN and Bernstein and Kramer would get together twice, first on January 3 or 4 in Washington, D.C., and then again in New York several days later.

A FUSE WAS LIT. On January 10, not long after David Kramer's second meeting with Carl Bernstein, CNN ran an on-air segment about the dossier. It disclosed that James Comey had given Donald Trump a two-page memo summarizing information from a document that claimed that the Kremlin had "compromising" sexual material on him and alleged that his staffers had colluded with Moscow during the campaign. The CNN report didn't mention Steele's name or Fusion GPS. But when it aired, David Kramer was inside the McCain Institute where a television set was tuned to CNN. "I believe my words were, 'Holy Shit,'" Kramer recalled.

The editor of BuzzFeed, Ben Smith, made a quick decision. Reporters at the website were still trying to chase down the dossier but Smith decided he wasn't going to get beat. On the evening of January 10, not long after the CNN report, he pushed a button and the dossier's memos were posted on the website's home page.

The BuzzFeed article about the dossier didn't mention Steele or Fusion GPS. It was loaded with all kinds of caveats and stated that the media organization hadn't been able to verify material in the dossier. Smith defended his action by saying the dossier was fair game because it was a government document, but many editors and reporters considered his decision to publish Steele's reports verbatim an act of journalistic malpractice. The reaction of Simpson, Steele, Peter Fritsch, and David Kramer was more immediate, because they feared that the dossier might contain enough clues for Russian intelligence to track down Steele's sources.

Simpson called a BuzzFeed reporter and screamed at him to delete the documents. "Take those fucking reports down right now!" he yelled. "You are going to get people killed!" Buzzfeed removed some potentially identifying material from the version of the dossier on its website.

IN PROMOTING THE DOSSIER to journalists, Simpson and Fritsch had laid down a condition—the names of Fusion GPS or Steele couldn't appear in any articles produced by journalists with whom they met. News organizations who had agreed to those deals now found themselves hemmed in, but with the Steele dossier public, the outing soon began.

On January 11, a day after BuzzFeed posted the dossier, Glenn Simpson learned that *The Wall Street Journal* planned to publish an article identifying Steele as its author. Simpson contacted editors at his old newspaper and pleaded with them not to run the article. But that approach failed and, the next day, it was Fusion GPS's turn in the spotlight. A *New York Times* reporter, Scott Shane, called Simpson to let him know that the paper was about to post an article disclosing his firm's involvement in the dossier. Simpson was irate and contacted Matt Purdy, the *Times* investigative editor he had met at the Democratic convention, to argue that the paper was reneging on its agreement not to identify Fusion GPS. Purdy replied that the role of Simpson's firm in the dossier story was widely-known.

GLENN SIMPSON FELT *The New York Times* had burned him. In a sense, he may have been right. Still, Simpson, Peter Fritsch, and Christopher Steele had spent months enjoying the comforts of anonymity while setting fires. If they got torched in the end that was an occupational hazard of their well-paid profession. As for Steele, he may have sensed before BuzzFeed posted his dossier that the jig was up. On January 5, 2017, right after he met with Ken Bensinger in London, he took a curious step. He deleted all his files about the dossier from his computer along with his email exchanges with Fusion GPS.

WHEN CHRISTOPHER STEELE'S MEMOS were posted on BuzzFeed's website, some of his competitors in the business of private spying were shocked. Professional jealousies may have played a factor in their assessments of the quality of his work. But for the most part, other operatives were uniform in their judgments: Steele's reports struck them as unprofessional and sloppy.

For starters, they pointed to his spelling of Alfa Bank in one memo as "Alpha" Bank. Steele said that Russian experts often translated the Cyrillic character for "f" into "ph" but even so, that criticism was but one of many. Some observed that his memos were written in differing styles, as though different people had authored them. Steele also had the habit, one corporate investigator pointed out, of citing what his sources believed other people were thinking. "It read like creative writing," that operative said, saying he was shocked to learn that a former MI6 agent had written the reports. One London-based corporate investigator, Andrew Wordsworth, who had worked with Glenn Simpson on the Prevezon case, was so appalled when he saw the dossier that he emailed Simpson, unaware that he had commissioned Steele to produce it. "Who wrote this 'shit'?" Wordsworth asked. He didn't hear back.

FOR JOURNALISTS, THE DOSSIER'S release by BuzzFeed wasn't the end of the Trump/Russia story. Instead, it was only the beginning. Later on, it would be hard to remember how cautiously media organizations had treated Christopher Steele's reports during the 2016 campaign. Prior to Election Day, only two outlets who spoke with him—*Yahoo! News* and *Mother Jones*—ran articles related to the dossier. *The New York Times, The Washington Post, The New*

Yorker, and other major outlets to whom Fusion GPS had shopped the memos hadn't touched them.

But now, with Steele's dossier public, the chase was on and it was time for journalists to take sides. In many cases, those choices were predictable. Even before Donald Trump's election, media rage machines on either side of America's growing political divide—Fox News on the right and MSNBC on the left—had been framing events through an ideological lens. But for journalists who considered themselves to be part of the mainstream media, the lure of the dossier and the seeming answers it provided to Trump's improbable victory proved irresistible.

If Glenn Simpson had ever fantasized about being a journalistic Pied Piper, BuzzFeed's decision to publish the dossier made that dream come true. And some reporters no doubt fantasized as well that the dossier would lead them to the kind of journalistic glory that greeted Carl Bernstein and Bob Woodward when they broke the Watergate story. As it turned out, the former reporting partners had very differing takes on Steele's memos. Bernstein liked them. Woodward called the dossier a "garbage document."

The end result was a media clusterfuck of epic proportions, one that was the consequence of the long-metastasizing relationship between private spies and journalists. In chasing the Trump/Russia collusion story, many reporters, including very good ones, never subjected Fusion GPS, Christopher Steele, or the dossier's claims to the same type of scrutiny they applied to other stories they pursued. Reporters who advocated "accountability" journalism gave in to the temptations of "access" journalism and the pitfalls it entailed. Plenty of reporters were skeptical of the dossier but they hesitated to dismiss it, because they didn't want to look like they were carrying water for Trump or his cronies.

Donald Trump's election would mark the start of an unparalleled era of governmental corruption, political cowardice, and social upheaval. Russia's efforts to meddle in U.S. politics were a real threat to America, one that Trump seemed to care little about. But there were other serious threats emanating directly from the Trump administration: attacks on science, contempt for the rule of law, campaigns to silence critics and intimidate career public servants.

Had Steele, Glenn Simpson, and Peter Fritsch stuck with the dossier's more modest finding that the Kremlin had tried to influence the 2016 campaign, the former MI6 spy's reports would have quickly faded from view. But they had overreached by weaving a larger web of conspiracy that created a litmus test, one that would ultimately benefit Donald Trump. Each time a claim in the dossier was shown to be wrong, Trump and his allies were handed a weapon they would use to damage the credibility of journalism at a time when it was needed most.

"It very much became almost a dispositive test," a former top FBI agent, Peter Strzok, said in a later interview published in *The Atlantic*. "Here's what's alleged to have happened, and if it happened, boy, it's horrible—we've got a traitor in the White House. But if it isn't true, well, then everything is fine. It framed the debate in a way that was actually harmful."

CHAPTER 12

TROJAN WARS

TORONTO, 2017

It was inevitable that Black Cube, given its arrogance and sleazy tactics, would eventually fuck with the wrong people. And for the Israeli firm, those people turned out to be a group of cybersecurity experts at Citizen Lab, an organization affiliated with the University of Toronto.

Founded in 2001, Citizen Lab was a Global Witness–style watchdog for the digital age, investigating how governments worldwide used new technologies to spy on their citizens and political opponents. One of its first reports to gain attention disclosed how the Chinese government was using cyber spying to monitor Tibet's Buddhist community, including its spiritual leader, the Dalai Lama. Another investigation looked at how the Mexican government was spying on its political opponents.

As part of its mission, Citizen Lab also tracked a hidden industry that produced advanced malware capable of infecting the cellphones of unsuspecting users and monitoring their conversations, emails, texts, and physical location. One kingpin of that dark domain was NSO Group, an Israeli company that produced a spy-

ware program known commonly as Pegasus. The government of
Saudi Arabia had used Pegasus to infect the phones of its critics,
including friends of Jamal Khashoggi, the opinion writer for *The
Washington Post* who was murdered and dismembered by Saudi
agents inside the country's embassy in Istanbul.

CITIZEN LAB'S ENCOUNTER WITH Black Cube started when a researcher with the group received an email from a man named
Gary Bowman, who claimed he was an executive at a Madrid-
based technology company called Flame-Tech. Bowman was eager
to speak with the researcher, Bahr Abdul Razzak, who emigrated
to Canada from Syria to escape the conflict there, about refugee-
related initiatives that Flame-Tech wanted to support. They met at
a Toronto hotel and, before long, Bowman started asking Abdul
Razzak questions about Citizen Lab's investigations into NSO,
the maker of Pegasus. He also wanted to know his attitudes toward Israel. "Do you write about it because it's an Israeli company?" Bowman asked. "Do you hate Israel?"

Abdul Razzak was shaken by the encounter and alerted his
Citizen Lab colleagues, one of whom suspected it was a Black
Cube operation. That researcher, John Scott-Railton, had joined
the group in 2012, when he was twenty-nine. Scott-Railton, who
had reddish hair and a closely trimmed beard, liked to speak to
journalists and he reached out to a cybersecurity reporter at the
Associated Press. He told the journalist, Raphael Satter, about
Bowman and Flame-Tech. Satter, who was then based in London, went to the British Library and pored through directories,
searching for listings in them for the businessman and his company, but didn't find any. Another AP reporter went to the address
in Madrid listed by Flame-Tech as the location of its offices. On
its website, the company claimed to occupy 20,000 square feet of

space on the building's seventeenth floor. But there was nothing to suggest it had ever operated out of the building or that Flame-Tech even existed.

Satter was about to publish an article about the episode when Scott-Railton urged him to hold off—he had just gotten an approach from another person who he suspected was a Black Cube operative. That man, who called himself Michel Lambert, had contacted Scott-Railton saying that he wanted to discuss a job opening at his Paris-based consulting firm, CPW Consulting. His company, he said, specialized in advising clients in Africa about agricultural technologies and had seen Scott-Railton's research about aerial mapping techniques and found it fascinating. "I've read your work and would like to discuss it with you to see where we could possibly apply it," Lambert said.

Scott-Railton smelled a rat. Long before he joined Citizen Lab, he had gone to Senegal to study how residents of Dakar living in shantytowns were adapting to increased flooding caused by climate change. The ideal way to track those changes was from the air, but Scott-Railton, who was pursuing a doctoral degree, didn't have the money to buy satellite images and, at the time, aerial drones were very expensive. So he came up with a fix. He tethered remote-controlled cameras to large kites that were sent aloft. He was so pleased with the quality of the resulting photographs that he posted them on his personal website. "I fly robot cameras on kites," he wrote.

But by the time Michel Lambert contacted him, that research was nearly a decade old and the use of aerial drones, whose prices had plummeted, was now ubiquitous. When Lambert asked Scott-Railton to meet with him in Toronto, the researcher made up his own story, saying he was moving to New York and was going there to search for an apartment. Lambert offered to buy him lunch at the Peninsula Hotel, one of Manhattan's most elegant. Scott-Railton accepted the invitation. Then he hung up the phone and called Raphael Satter, the AP reporter.

SINCE ITS START, CITIZEN Lab had focused on government-sponsored cyber-spying. But in time, Scott-Railton and a colleague would stumble over another sector where the use of hacking and malware was booming—the private spying industry.

Corporate investigative firms all deny they hack because it is illegal. But since the days of "Operation Hellenic" in 2007, the industry's appetite for cyber-surveillance and digital snooping has blossomed. In 2010, two Israeli private investigators hired by an associate of Oleg Deripaska, the Russian aluminum kingpin, were convicted of wiretapping one of the oligarch's former business partners with whom he was involved in a legal dispute. A year later, in 2011, the hacking scandal at Rupert Murdoch's *News of the World* erupted and, that same year, the head of security at a French energy firm was convicted of hiring a private operative to burrow into the email of Greenpeace, the activist group. Then, there were the incidents involving ENRC, the Kazakh-connected mining company, and its allied company, IMR, and, in 2015, the FBI arrested dozens of people who were charged with selling hacking services to private investigators.

Hacking and cyber-spying was growing more common among hired spies because experts who had learned their skill while working for government intelligence agencies or the military were now selling them to private customers. In addition, once-costly electronic surveillance tools developed for use by intelligence agencies or the police had become cheaper and widely available. One of those gadgets was a piece of eavesdropping equipment known as an "IMSI catcher." It was essentially a miniature cellphone tower and an operative, by standing between their target and a real cell tower, could use the device to intercept calls. Rogue cellphone apps were also used to monitor subjects. Ronan Farrow, the *New Yorker* reporter, later described how he kept receiving weather alerts on his cellphone while working on stories about Harvey Weinstein.

To monitor their targets, some operatives-for-hire also piggybacked on a system used by bounty hunters to find fugitives. Several of the major cellphone carriers sold real-time data about the location of a customer's cellphone to licensed bounty hunters to help them locate their quarry. And for their part, bounty hunters did a brisk side business taking orders from private operatives for cell numbers they wanted to track. A reporter for a news website, *Motherboard*, exposed the practice and described how he had paid three hundred dollars to a bounty hunter to track his phone.

THE TRAIL THAT EVENTUALLY led John Scott-Railton to discover the scope of cyber-spying by private operatives began in 2017 when a journalist contacted Citizen Lab about a suspicious-looking email. The message looked like a security alert from Google, a common ploy used by hackers to get a target to divulge their email password and thereby gain access to their account.

When Scott-Railton and a colleague, Adam Hulcoop, examined the email, they discovered it contained a shortened hyperlink, or URL, that hackers had created from a do-it-yourself program on the internet. Hulcoop, who worked as a computer security expert at a Canadian financial institution, was able to unshorten the URL and saw that the full version of it contained the email address of the person targeted. He and Scott-Railton then discovered a code that the hackers had used to create the shortened links, and using it as a starting point, they were able to find thousands of other phishing emails sent out by the same hackers. The locations and occupations of those targeted were so diverse—they included, among others, journalists, lawmakers, attorneys, activists, and politicians in numerous countries—it was clear that a single government couldn't be behind the campaign. Instead, the attacks looked like the handiwork of a hacking-for-hire firm that was carrying out assignments for a large number of private clients.

IN 2017, JOHN SCOTT-RAILTON would meet one of those targets. The Citizen Lab researcher was at a conference in Virginia when another attendee introduced himself. The man's name seemed very familiar to Scott-Railton and then he realized why—his name was on the list of people targeted by hackers. Soon after the meeting, Scott-Railton contacted the man, who was involved with an environmental group taking part in a campaign called "Exxon Knew," that accused Exxon Mobil of hiding its knowledge of how fossil fuels were driving climate change, a claim the company denied. Scott-Railton was soon introduced to other activists involved in the campaign who said they had received emails from suspected hackers as well as fake videos and Dropbox messages. Some activists had clicked on those links and unwittingly given the hackers access to their computers.

Scott-Railton and Adam Hulcoop were certain that the hacking-for-hire operation was based in India but proving it and figuring out who had ordered the hacking and benefited from it required a different set of skills. So he and some environmentalists met with people who might be able to untangle those connections—prosecutors with the U.S. Justice Department.

JOHN SCOTT-RAILTON WAS ALREADY seated at a table in the dining room of the Peninsula Hotel in Manhattan when Michel Lambert, the suspected Black Cube operative, arrived. Stocky and balding, Lambert appeared to be in his sixties and was dressed in a brown suit, white shirt, and tie. He and Scott-Railton had both brought backup with them. Two local private detectives whom Black Cube had used to shadow reporters investigating Harvey Weinstein had staked out the hotel. Raphael Satter, the AP reporter, sat at a table at the far end of the restaurant along with a videographer for the wire service, and an AP photographer loitered in the hotel's lobby.

Over the previous week, Scott-Railton and Satter had researched Lambert and his firm, CPW Consulting. Like Gary Bowman and Flame-Tech, they quickly proved to be online ghosts. The supposed consultant had a LinkedIn profile with five hundred connections but Satter couldn't find any real-world references to him or his firm in directories. For his part, Scott-Railton observed Black Cube's digital producers build the website of CPW Consulting. At one point, an ad appeared on the firm's LinkedIn page looking for a digital mapping specialist. The opening was genuine. It just wasn't for a job at CPW Consulting. Scott-Railton ran phrases from the posting through Google and discovered that a local British housing authority had posted the job on its website. Someone at Black Cube, it appeared, had copied the ad and, after making slight modifications, pasted it onto CPW's page.

Scott-Railton had made other preparations for the meeting. To embellish his own cover story, he had posted online advertisements seeking an apartment in Brooklyn and contacted brokers who were sending him rental listings. He had also jury-rigged a miniature video camera that was now concealed behind his tie and he had audio recorders in his pockets.

THE LUNCHEON WENT ON for about ninety minutes. John Scott-Railton and Michel Lambert both spoke in French and the supposed consultant, who said he had grown up in Morocco, described how his work regularly took him to Africa. He came across to Scott-Railton as bumbling. As he spoke, he took index cards from his jacket pocket and kept referring to them as though they contained prompts. He also pointed a pen toward Scott-Railton that the researcher suspected contained a recording device.

Lambert probed for details about Citizen Lab, including where the group got its financial support and how it chose its targets. The answers that Scott-Railton gave him were vague but he told Lambert that the group was very interested in NSO, the maker of Pegasus, the spyware. As they chatted, Scott-Railton regularly checked his cellphone, telling Lambert that real estate brokers were sending him listings. In fact, he was exchanging text messages with Raphael Satter.

"You found an apartment?" Lambert asked him.

"Not yet," Scott-Railton replied. "And then there are a fair few scammers."

"Scammer?" Lambert responded, seemingly unfamiliar with the word.

"Scammers—I don't know how to say it in French—con artists," Scott-Railton said.

Both of them ordered crème brûlée for dessert. Meanwhile, Satter was getting anxious. He and the AP videographer lingered for ninety minutes over two glasses of wine and a half-dozen oysters while a waiter kept pressing them to order. The battery in the hidden microphone that Satter was wearing was draining and he texted Scott-Railton that they needed to move. Scott-Railton gave him the go-ahead and the AP reporter approached Lambert, who was now lingering over coffee. He identified himself as a reporter.

"I'd like to talk to you about your company," Satter said.

"I don't have to talk with you," Lambert responded.

"Actually, I think you'll want to," Satter replied. "My colleague has been visiting your company this morning and she said it is very strange but it doesn't exist."

AP reporters had gone to the supposed address of CPW Consulting in Paris and discovered it was an apartment building. None of the residents had ever heard of CPW Consulting and one of the AP journalists had taken a photograph of herself standing in front of the building. She was holding a sign that said, "Hello, Michel."

Lambert got up from the table and scrambled to avoid Satter and his cameraman. He found a door out of the restaurant and ran through it. One of the private detectives hired by Black Cube had been lingering at the restaurant's bar and wasn't in a position to see the confrontation. Then he got a call from his colleague who was in contact with Lambert. "What the fuck is going on," the other detective said. "Who are the guys with the cameras."

When that detective, Igor Ostrovskiy, looked up, he saw people running toward the stairs leading down to the hotel's lobby. He followed them out onto the street and found four people standing together there: Scott-Railton, Satter, and the two other AP staffers. He didn't know any of them and took a picture of the group with his cellphone camera and sent it to his fellow private detective. Then he sent a copy of the photograph to someone he had gotten to know and befriended, Ronan Farrow, the *New Yorker* reporter.

WITHIN THE INVESTIGATIVE INDUSTRY'S pecking order, Igor Ostrovskiy occupied one of its lower rungs. In his mid-thirties, the private detective was short and slightly overweight. He made his living on the type of run-of-the-mill case that big corporate intelligence firms wouldn't touch because they didn't pay enough. Typically, they involved divorces, suspect insurance, or worker's comp claims and the occasional missing person. Ostrovskiy spent lots of his days sitting in his car doing stakeouts and eating bad food.

There was something else that set Ostrovskiy apart from many of the well-dressed operatives at K2 Intelligence, Kroll, and other big-name corporate intelligence firms. He had a sense of ethics and he had gotten upset after learning about Black Cube's efforts to kill journalistic exposés of Harvey Weinstein.

When Ostrovskiy first got involved in the Weinstein case he didn't know what it was about and had never heard about Black Cube. Instead, another New York–area private investigator hired him to help on stakeouts. That private eye hinted to Ostrovskiy that their client was an Israeli firm and soon they were shadowing a *New York Times* reporter, Jodi Kantor, and Ronan Farrow.

Ostrovskiy would only learn the reason why when he read Farrow's piece in *The New Yorker* that exposed Black Cube's tactics. Big-time lawyers at David Boies's firm weren't seemingly troubled by them but Ostrovskiy was appalled. He had been nine years old when his parents emigrated to the United States from Ukraine, and had embraced the values of his new homeland. Ostrovskiy was so upset by the idea that a private intelligence firm located outside the United States was trying to subvert the free press that he contacted Farrow and began providing him with information about his assignments for Black Cube.

IT DIDN'T TAKE LONG after the AP published Michel Lambert's photograph for his real name to emerge. On the day after the Peninsula Hotel meeting, a television station in Israel and *The New York Times* identified him as a retired Israeli security official named Aharon Almog-Assoulin and it became clear that he was another cast member in Black Cube's cabinet of goons. A lawyer in Canada representing an investment firm engaged in a lawsuit there against a rival said a man closely resembling Almog-Assoulin had approached him, trying to get information about the case. He had given his name as Victor Petrov and said he worked for a firm called KWE Consulting, which proved to be a phantom.

Ostrovskiy would later say that he had seen Almog-Assoulin before the Peninsula Hotel while working on other Black Cube–related assignments. He hadn't known his name but in each case, he had served as backup for Almog-Assoulin during meetings at hotels or restaurants to which the Black Cube operative had lured a target.

Ostrovskiy said his instructions were always the same. He was to seat himself behind Almog-Assoulin and take photographs that showed the target's face but didn't reveal any of the operative's features. He then sent the pictures to his fellow private investigator who was working as a contractor for Black Cube. Afterward, his instructions were to delete the images from his camera's memory card. The whole point of the exercise was clear to him. "I knew they were using these kinds of pictures to embarrass or discredit a target," he said.

After the incident at the Peninsula Hotel, Ostrovskiy was told by the private investigator who hired him to work for Black Cube that the Israeli firm wanted them to take a polygraph test. It was trying to figure out who had tipped off John Scott-Railton or Raphael Satter about the operation at the hotel. Ostrovskiy refused to take the test. He would rather sit in his car on freezing days trying to catch insurance cheats than have anything to do with Black Cube again.

CHAPTER 13

ROCK STARS

Glenn Simpson was swarmed by reporters upon his arrival in 2017 at a conference in Aspen, all of them eager to score his first public interview about the dossier. Better yet, some were hoping that Simpson would set up a meeting for them with Christopher Steele. "Everyone wanted Glenn," Rhonda Schwartz, the ABC News producer, said. "He was like a rock star."

Schwartz watched as two NBC News correspondents, Andrea Mitchell and Lester Holt, snagged Simpson for dinner. She was so worried that NBC News would get the scoop that she called her reporting partner, Brian Ross, to figure out how they should respond. As it turned out, there wasn't really a need to do anything. Simpson had apparently set a ground rule for landing an interview with him. A year had passed since Fusion GPS had worked on behalf of the Russian-owned real estate firm, Prevezon Holdings, but Simpson's animus for Bill Browder apparently remained un-requited. "In order to get Glenn, you first had to do a hit piece on Browder," said Schwartz, adding that she and Ross weren't inter-ested in taking up Simpson on that deal.

WITH HILLARY CLINTON'S DEFEAT, payments from the Democratic Party to Fusion GPS had ended. But Glenn Simpson and Peter Fritsch found a new and profitable way to fund their ongoing investigations into Donald Trump and Russia. Since their firm was a for-profit business, they couldn't go out and solicit contributions for it. So they did the next-best thing. They helped form a nonprofit foundation that accepted millions of dollars from donors who despised Trump and then funneled money to Fusion GPS and Christopher Steele.

The stated mission of the group, called the Democracy Integrity Project, was to sponsor research into Russian meddling in elections worldwide. Simpson and Fritsch wanted to give it the veneer of a nonpartisan, journalistic-style organization by having a high-profile reporter help lead it. That proved to be a nonstarter. To keep the names of its donors secret, the Democracy Integrity Project was set up under a provision of the U.S. tax code that legally allows nonprofits to conceal the names of supporters. Organizations of all political stripes take advantage of the provision to mask their donors' identities but many journalists, who refer to such groups as "dark money" organizations, find them anathema.

Simpson, however, soon found someone to head the Democracy Integrity Project who had star power with both the news media and Democratic givers. He was a former U.S. Senate intelligence committee staffer, Daniel S. Jones, who had gained fame in 2014 for his role in bringing to light the CIA's use of torture to interrogate suspects during the "War on Terror."

At the time, Jones, who had earlier worked at the FBI as an analyst, was serving as the chief investigator for the Senate Select Intelligence Committee while it was gathering evidence on techniques used by CIA interrogators at Guantanamo Bay. The spy agency was bitterly fighting Jones's efforts to make that dark chapter of its history public, but he refused to relent. The story of his battle with the CIA became the subject of a 2019 movie, *The Report*, in which Jones was portrayed by the actor Adam Driver. At showings of the film, Jones, who had an intense, driven manner, was lauded as a champion of transparency and he received a standing ovation following one screening from an audience whose members included the documentary filmmaker Michael Moore.

In early 2017, Simpson and Jones were out making fundraising pitches to deep-pocketed Democratic donors in Silicon Valley, Hollywood, and New York who hoped that Fusion GPS and Christopher Steele, given time and money, could nail down the case against Donald Trump. According to reports in *The Daily Caller* and other conservative publications, the group's donors included George Soros, the billionaire investor; Rob Reiner, the actor and director; and a group tied to Tom Steyer, a California businessman who sought the 2020 Democratic presidential nomination.

In 2017, its first year of operation, the Democracy Integrity Project pulled in more than $7 million, a stunning haul. About half that sum—or $3.3 million—was paid out in fees to a limited liability company linked to Fusion GPS. That figure represented nearly three times the amount that the firm had received from Hillary Clinton's campaign. An entity tied to Christopher Steele got $252,000, or about twice as much as it had received during the 2016 campaign from Fusion GPS. Some of the group's money went to a Fusion GPS contractor in England, where the firm was generating news articles seeking to tie supporters of the Brexit campaign to the Kremlin. As for Dan Jones, his starting salary in 2017 was $381,000.

Soon after the Democracy Integrity Project started, Simpson brought Jones with him to the annual journalism conference hosted by Lowell Bergman, the former *60 Minutes* producer, at the University of California, Berkeley. Simpson introduced Jones to reporters there as a future point of contact on the Trump/Russia story. "He is going to take this into another dimension," Simpson told one of them.

The Democracy Integrity Project would send a daily news summary to journalists and congressional staffers highlighting published pieces about the Trump/Russia controversy and worked with reporters to generate more coverage. In early 2017, Jones wrote an email to a Washington lawyer and lobbyist who was close to Christopher Steele and Oleg Deripaska, the Russian oligarch. The email linked to a Reuters piece in which the news organization reported that a review of units in a Trump-branded building in South Florida had found that sixty-three of them were purchased by individuals with Russian passports or addresses. "Our team helped with this," Jones wrote.

Thanks to the dossier, it was a seller's market for Trump/Russia stories. In early 2017, *The Wall Street Journal* broke a story reporting that Sergei Millian, the self-styled real estate broker, had been a key source for some of the dossier's "most explosive parts." In the article, a *Journal* reporter, Mark Maremont, wrote that a "person familiar with the matter" said it was Millian who had provided information about the "pee tape" and Trump's ties with Moscow, though the broker wasn't aware while talking with Steele's collector that he was speaking with a hired spy. Both ABC News and *The Washington Post* soon ran similar pieces describing Millian as a critical source for the dossier. "Trump had a long-standing relationship with Russian officials, Millian told an associate, and those officials were now feeding Trump damaging information about his Democratic opponent, Hillary Clinton," reported the *Post*, which identified Millian as "Source D" or "Source E" in Steele's memos, or the person whom the ex-MI6 agent described as having "direct or indirect" knowledge of the "pee tape."

From a journalistic perspective, the *Journal*'s scoop was particularly interesting because it cited only one unnamed person as the source of the information about Millian and newspapers typically require a reporter to have two sources. But it seems that the material in that article and the ones following it apparently emanated from Simpson and Fritsch, who were eager to generate coverage of the dossier. Two reporters said that the Fusion GPS operatives told them about Steele's key and how it identified Sergei Millian as a source. (Millian has denied being a source for any information in the dossier.)

IN MARCH 2017, JAMES Comey, the FBI director, disclosed at a congressional hearing that the bureau had been conducting a counterintelligence investigation into the Trump campaign. At the same hearing, Democratic congressman Adam Schiff of California, a leading Trump critic, read aloud from one of Steele's memos that reported that officials of Russia's biggest oil company had dangled a financial stake in front of Carter Page as a consulting fee (Page has rejected the assertion). Soon afterward, Trump fired Comey, setting the stage for the appointment of Robert Mueller III as an independent federal prosecutor to investigate Russia's interference in the 2016 election.

Some journalists reported that U.S. authorities had found evidence to support Steele's critical claims. "They now have specific concrete and collaborative evidence of collusion," wrote Luke Harding, a reporter for *The Guardian* who was close to Steele and Simpson. Reporters and commentators on CNN made similar claims and other journalists would later say that a reporter for *Politico*, Natasha Bertrand, who was also an MSNBC commentator, became a regular outlet for material that Fusion GPS was promoting. (Bertrand would reject that suggestion.)

But few things seemed to make Glenn Simpson happier than

seeing Rachel Maddow, the fiery MSNBC host, talk about the dossier. And if that was the case, he was a very happy man. Maddow constantly talked about Trump, Russia, and the dossier, spinning out elliptical commentaries.

"We have a special report for you tonight on a story we've been working on for a very long time," Maddow said at the start of one 2017 show that was devoted entirely to Steele's reports. "We are going to step back and look at the thirty-five-page Trump Russia dossier. And depending on which way the news is blowing, the allegations in this document can sound outlandish or they can sound freakishly spot on."

ALONG WITH ESTABLISHED OUTLETS for Trump/Russia stories, Fusion GPS was eager to broaden the audience and invited news organizations to its Washington offices for briefings. Simpson started one by describing himself as the kind of journalist who had seen it all and whose reporting experiences at *The Wall Street Journal* had led him to understand how the Russians had corrupted Trump. He then ticked off a laundry list of people that Fusion GPS was still tracking and whom he thought that the journalists attending the briefing should consider as ripe targets for investigation. That group included, among others, Paul Manafort, Carter Page, Sergei Millian, and Rudolph Giuliani. One reporter present asked Simpson to name the person upon whom he would focus his energies if he had one choice. "Dude, that changes from day to day," Simpson replied.

To some journalists, Simpson seemed to be pushing Trump-related stories that felt old and stale because they had already gotten a lot of media coverage. At the end of one meeting, a Fusion GPS executive pulled aside an editor from *Vice News* to privately pitch him on what he thought could be a hot new story.

During the 2016 presidential campaign, there had been rumors about a secret tape containing outtakes of racist remarks that Trump had made during the filming of episodes of *The Apprentice*, the reality show on which he had starred. That tape never materialized but the Fusion GPS executive said he had heard rumors about another incriminating Trump video, one supposedly showing Trump viciously striking his wife, Melania, inside an elevator. The operative added that, according to rumors, a security camera had captured the scene and that Rupert Murdoch, the media mogul and a Trump supporter, had supposedly purchased the tape at a black-market auction as part of a catch-and-kill operation to make sure it never emerged. The *Vice News* editor listened and then left. Reporters for *Vice*, *The New York Times,* and other media outlets had been hunting for the rumored tape before concluding it didn't exist.

ALONG WITH ATTENTION AND money, Glenn Simpson, Peter Fritsch, and Christopher Steele were also getting plenty of blowback. Donald Trump and his political and media allies were hellbent on destroying them. They were also eager to use questions over the dossier's accuracy as a battering ram to demolish findings by U.S. intelligence agencies that Moscow had tried to meddle in the 2016 campaign. Trump described Steele's memos as "garbage" and "fake news" and commentators on Fox News called them "claptrap," "Russian lies," and a product of the "deep state" conspiracy to diminish the validity of Trump's victory.

Opening one show, Sean Hannity, the popular Fox News host, told his viewers about the recent disclosure that the wife of Bruce Ohr, the Justice Department official to whom Simpson had given a thumb drive with the dossier, once worked at Fusion GPS. "Fusion GPS, the firm behind the Clinton bought and paid for fake

Russian dossier, well guess what, they admit the wife of a recently demoted DOJ official worked at the company on Trump opposition research," Hannity declared. "This can't be in a novel anymore, you can't make it up."

REPUBLICAN LAWMAKERS ALSO CAME after Fusion GPS and Christopher Steele with a vengeance. Among the most rabid was Congressman Devin Nunes, a Republican from California, who dispatched two aides to London to try to spring a surprise visit on Steele at the offices of Orbis Business Intelligence. Republican senator Charles Grassley of Iowa, another Trump ally, pounced on a complaint that Bill Browder filed a year earlier where he accused Glenn Simpson, Natalia Veselnitskaya, the Russian lawyer for Prevezon, and other people of violating federal law by failing to register as foreign lobbyists while working to undermine the Global Magnitsky Act.

Browder's complaint went nowhere but three operatives from Fusion GPS—Simpson, Peter Fritsch, and Tom Catan, another ex–*Wall Street Journal* reporter—were summoned to testify before Congress about the investigative firm, its work on behalf of Prevezon Holdings, and the dossier. Fritsch and Catan invoked their Fifth Amendment right not to testify. Simpson didn't and took center stage.

It was his public coming-out. Stories about him or Fusion GPS appeared in *The New York Times* and *The New Yorker*. "Mr. Simpson, a 53-year-old Wall Street Journal veteran-turned-master of high-dollar research, has arrived at the biggest story of either of his careers, lurching to the center of the Russia-tinged scandal that clouds the presidency," the *Times* article stated.

Simpson showed a talent during his congressional appearances for casting himself in the best possible light. He responded to lawmakers' questions by downplaying the role of Fusion GPS in the Prevezon case and claimed that he and Bill Browder, despite all his efforts to undermine him, were at heart political blood brothers. "Sergei Magnitsky was killed in prison by neglect if not worse," Simpson said.

At the time of Simpson's congressional appearances, the news about the Trump Tower meeting involving Natalia Veselnitskaya and Donald Trump Jr. had just broken. Those articles also disclosed that the Russian lawyer had given a memo to the candidate's son that Simpson had prepared while working on behalf of Prevezon. One Republican staffer kept pressing Simpson during his Congressional testimony about his work for Prevezon, asking him whether he felt in hindsight that Veselnitskaya and others had played him to advance the Kremlin's interests. Simpson replied that he wasn't happy about what had happened but added that there wasn't much he could have done about it. "I am in the information business so when people commission research from me, it becomes their property," he said. "If they decide to go and use it for something else, I mean that is just beyond my control."

SIMPSON AND FRITSCH SOON found another forum to make their case—the opinion pages of *The New York Times*. They depicted themselves in one op-ed piece they both wrote as victims of a character assassination campaign by Trump's allies. In a second piece that carried only Fritsch's byline he complained that Republican lawmakers had forced Simpson, despite hours of earlier testimony, to make a "perp-walk" before Fox News cameras to create a false impression of guilt after he invoked his Fifth Amendment right not to testify further. "We have nothing to hide," Fritsch wrote. "But [we] will not be drawn into a rigged game based on a playbook borrowed from the McCarthy era."

Meanwhile, lawsuits were starting to pile up against Fusion GPS and Orbis Business Intelligence. In 2017, the three founders of Alfa Bank sued Fusion GPS and Orbis, charging that the information Steele had reported about them was defamatory. Separately, BuzzFeed and Steele's firm were sued by the Russian owner of several internet service providers who was described in a final memo Steele wrote after the 2016 election as a participant in Moscow's clandestine cyber attack against Hillary Clinton's campaign.

The lawsuit against BuzzFeed was dismissed. And to defend themselves against the Alfa Bank–connected actions, Fusion GPS and Orbis Business Intelligence, which both denied any wrongdoing, employed the same strategy that Simpson had used in 2013 when Frank VanderSloot, the Mitt Romney donor, came after him. Both firms invoked anti-SLAPP statutes as protection and moved to have the oligarchs' claims thrown out. Orbis prevailed in the United States but Fusion GPS wasn't so lucky and a judge allowed the claims against it to go forward. Still, Steele's firm wasn't in the clear. It continued to face dossier-related claims in London from both the Alfa Bank–connected oligarchs and the owner of the internet services companies.

FROM A LEGAL PERSPECTIVE, the lawsuits against Fusion GPS, Orbis Business Intelligence, and BuzzFeed differed. At their core, they were connected by a common thread—the credibility of Christopher Steele's reports. In the aftermath of the dossier's appearance, the former spy had avoided the press. But with the lawsuits against his firm and Fusion GPS moving forward, he started speaking with journalists deemed friendly either directly or through intermediaries. Glenn Simpson and other of his allies served as Steele's gatekeepers, approving reporters who wanted access. The results were predictable. Journalists wandered into an echo chamber and the portraits of Steele they produced portrayed his career, his investigative prowess, and his motives in a glowing light.

One of those allowed into Steele's inner sanctum was Jane Mayer of *The New Yorker*. She didn't interview Steele but spoke with his partner, Christopher Burrows, and other friends of the ex-MI6 agent. Mayer was considered one of the best investigative reporters in the United States and for good reason. She was a well-sourced, relentless digger who had examined topics as diverse as the political infighting surrounding the nomination of Clarence Thomas to the U.S. Supreme Court to the war on terror to the secret funding of antiregulatory campaigns. But in the view of some reporters, her profile of Steele wasn't her finest hour. She noted the morally ambiguous world that Steele now inhabited as a private operative and reported that Oleg Deripaska's lawyer had employed him. But she didn't scrutinize Steele's relationship to the oligarch with her usual zeal.

Two early books about the Trump/Russia controversy which delved into Moscow's efforts to influence the 2016 campaign cast Steele as a private sleuth turned public hero. One of them, titled *Russian Roulette*, was written by Michael Isikoff of *Yahoo! News* and David Corn of *Mother Jones*. The other book, titled *Collusion*, was written by Luke Harding, the *Guardian* reporter. Each book contained highlight reels from Steele's career, including what the authors described as his pivotal role in the FBI's investigation of bribery at FIFA, the organization governing soccer. "Steele discovered that FIFA corruption was global," Harding wrote in *Collusion*. "It was a stunning conspiracy." In their book, Isikoff and Corn sounded a similar note. "Steele's reports helped spur a wideranging, years-long investigation that ultimately led to multiple indictments against FIFA figures by federal prosecutors," they wrote.

Steele did pass along some FIFA-related information to the FBI. But his contributions, it emerged, differed from those descriptions. A Justice Department review of Steele's interactions with the FBI found that none of the material he provided to U.S. officials was used in any indictments against FIFA executives. That doesn't mean he wasn't helpful. But Steele basically acted as a middleman who connected the FBI with a British reporter, Andrew Jennings, who had broken the FIFA story.

FOR ANY JOURNALIST, EXAMINING the career of a spy or a private operative is never easy because their activities, by their nature, are secret. But in the hothouse environment of the Trump era, reporters writing about Christopher Steele seemed to skip over a fundamental question that ought to have been front and center: How the hell did he know things about Donald Trump and Russia that the CIA and MI6 didn't?

By 2016, Steele had been out of the game for years. He hadn't been to Russia for two decades and had left MI6 seven years earlier. Steele, like every other ex-spook in the private intelligence industry, was selling himself to private clients based on his past, the knowledge he had gained, and his network of informants. But Steele's sources weren't carryovers from his time as an MI6 agent but rather paid informants whom he had started using since becoming a private operative.

There are many things about which hired spies will disagree but virtually all of them will say that dealing with paid informants is one of the most treacherous parts of their business. Some informants will tell an operative what they think they want to hear and even well-meaning ones aren't all that reliable. Steele had insisted to Glenn Simpson, Peter Fritsch, and everyone else to whom he talked that his principal source, or "collector," for the dossier was an informant with contacts deep inside the Kremlin. That sounded great. It also sounded like a bit of a stretch. Even government intelligence agencies consider the recruitment of a single "high value" informant—a trusted source who is intimately close to the action—as an intelligence coup. The reason: it hardly ever happens.

WHEN BUZZFEED POSTED THE dossier, some news organizations reached out to writers of spy thrillers to get their take on Steele's memos. One of those authors was a former British politician, Rupert Allason, who wrote espionage novels under the pen name Nigel West. He and Steele had met after the former MI6 agent started his career as a private operative and Allason said that he helped connect an American espionage writer, Howard Blum, with Steele for the profile that Blum wrote about him for *Vanity Fair*.

"He's James Bond," Allason told NBC News when the dossier was released. "I actually introduced him to my wife as James Bond."

Allason's opinion of Steele, however, would change after a lawyer in Washington, D.C., with Republican Party connections hired him to do an analysis of the dossier. In fly-specking the memos, Allason noted plenty of problems, particularly with how Steele was attributing information to his sources. Steele cited eleven people as unnamed sources and assigned an alphabetic code name to seven of them. The trouble was that, as Allason saw it, Steele's code-named sources seemed to be suffering from the spy world equivalent of multiple personality disorder. According to Allason's analysis, one of Steele's sources—who was code-named Source "E"—was described in differing memos by the former MI6 spy as 1) someone who had access to the staff of the Ritz-Carlton hotel, as 2) someone who knew about the Kremlin's involvement with

WikiLeaks, and as 3) someone who knew about the use of Russian diplomatic missions for spying.

In the end, Allason came to believe that Steele or his sources were fabricating information. "There may be only one 'trusted compatriot' [*sic*] reporting on his private conversations," Allason wrote. "There is no indication of the reliability of this individual."

AFTER ALLASON'S REPORT PUBLICLY emerged, Orbis Business Intelligence issued a statement dismissing it as a "politically motivated" piece of fiction and adding that Allason had no knowledge of Steele's sources. But in time, another troubling theory about the dossier would begin to gain traction. It went like this: Russian intelligence operatives, aware that Steele was collecting information about Trump and the Kremlin, fed disinformation to his sources that then got incorporated into the dossier.

One journalist who came to suspect that early on was Scott Shane, an experienced investigative reporter for *The New York Times*. It was Shane who had enraged Simpson by telling him that the paper decided to out Fusion GPS, though Simpson and Peter Fritsch soon forgave the paper and gave a copy of the dossier to a group of its reporters in Washington, D.C.

Shane decided to run down Steele's memo about Michael Cohen's supposed meeting in Prague with Kremlin operatives. To the reporter, the allegation about Cohen's Prague trip was intriguing because it, unlike other parts of the dossier, contained a high degree of specificity. The report cited the time of Cohen's meeting, the name of the Russian cultural organization where it supposedly took place, and the name of the Russian lawmaker with whom Cohen was said to have met. Shane knew, like every reporter who had chased the Prague story, that Donald Trump's tenure as president might be short-lived if the allegation was true.

Shane, who had bottle-thick eyeglasses and an easy, thoughtful manner, spoke Russian and had lived in Moscow for a time as a correspondent for *The Baltimore Sun*. He had a good sense of how Russian intelligence agencies operated and was familiar with the Kremlin-linked group where the Cohen meeting supposedly took place.

The organization was called Rossotrudnichestvo and it sponsored Russian cultural events abroad as well as student exchange programs. The group had offices worldwide and they doubled, as government embassies often do, as places from which spies operate undercover. In 2013, the head of Rossotrudnichestvo's office in Washington, D.C., had left the United States after he came under FBI investigation as a spy.

Shane and a *Times* colleague tried every trick they could think of to track Cohen's whereabouts at the time of the Prague meeting. They examined the social media accounts of the lawyer's children to see if they had made any mention of their father being abroad. When those efforts didn't provide an answer, Shane decided to call the offices of Rossotrudnichestvo in Prague to see if he could bluff someone there into giving up information. To his surprise, he was connected directly to its head, who denied knowing Cohen or anything about a meeting. "He said, 'I don't know who started this rumor,'" Shane later recalled.

Then another thought occurred to Shane. The details and description of Cohen's clandestine meeting seemed too perfect. He knew that Russian intelligence agencies closely monitored the activities of private operatives and tracked people who they suspected were their sources of information. Shane began to imagine that Russian intelligence agents, having picked up reports that Steele was sniffing around about Trump, had assembled a computerized map showing the sources used by Orbis Business Intelligence. Then all Kremlin agents had to do was tap some of those people and drip disinformation into Steele's network.

Shane hadn't written about his theory since he couldn't prove it. But about a year after the dossier's release, he and other *Times* reporters took part in a panel discussion at the International Spy Museum in Washington, D.C., to talk about the Trump/Russia story. "I want to throw a curveball out for people to contemplate," Shane said. "But it is something that has bothered me for the whole past year since the dossier became public."

He then wondered aloud about how the report about the Cohen Prague meeting found its way into the dossier. "Here is the question, if that is not true, who had made it up?" he asked. "And the thing that occurred to me was that this was not planted by some Clinton operative. This was probably concocted by Russian intelligence. . . . And when you think about Christopher Steele, the retired British intelligence agent calling his buddies who called their buddies in Moscow to get this information, the FSB is certainly capable of following those trails and polluting them with certain disinformation."

Shane went on to say that it was possible that Kremlin operatives, while focusing the brunt of their disinformation campaign against Hillary Clinton, might have hedged their bets by planting dirt about Trump and his associates.

A former CIA spy on the same panel suggested that Shane had let his imagination run wild. "I mean anything is possible, you are absolutely right," she said. Then she recited the mantra that believers in the dossier would repeat again and again in years ahead. "I think with the dossier, time is going to tell, how much of it checks out," the ex-spy said. "But keep in mind, have they disproved any of it? . . . I think not."

CHAPTER 14

EPISODE 1: "DOUBLE AGENT"

LONDON, 2019

When Mark Hollingsworth's downfall finally came there was a sense of poetry to it. For decades, Hollingsworth had presented himself as a journalist while misleading reporters and getting paid by private spying firms. But what made his last act so satisfying was that it involved a new and secret role he played for ENRC, the mining firm he had written about for years. He became a snitch for the company.

By 2019, the bilateral warfare between Britain's Serious Fraud Office and the Kazakh Trio was still under way. Government investigators were continuing their probe of bribery allegations against ENRC and an army of lawyers and private operatives working on behalf of the company were pursuing a scorched-earth policy to end the inquiry by delegitimizing it. Their basic claim was that the British government's investigation had been poisoned from the start because Neil Gerrard, the outside lawyer hired nearly a decade earlier to investigate whistleblower allegations, had illegally given its records to the Serious Fraud Office and that government officials had then leaked information to journalists and others. To try to find evidence to back its position, ENRC was now suing lots of people who had intersected with that investigation over the years.

One of its top targets remained Neil Gerrard. ENRC had long insisted that Gerrard gave its records to the Serious Fraud Office without its approval and that his real goal, rather than defending the company, had been to drum up publicity and more legal work for himself. Private spies working on behalf of ENRC tried for years to dig up dirt on him and, in 2012, they appeared to catch a break. An anonymous writer sent a letter to British authorities claiming he had been present at a gathering where a drunken Gerrard had supposedly boasted about his ability to fix cases at the Serious Fraud Office.

When that letter emerged, Mark Hollingsworth sprang into action. "Please send me anything you can find, especially on Gerrard's finances and country house," he wrote to a researcher. He then sent nearly identical reports about the lawyer to operatives at two firms he worked for as a contractor. One went to Charlie Carr, who was still at K2 Intelligence. The other report went to an executive at Alaco, the firm for which Hollingsworth reportedly posed as a journalist in Ireland.

"While this could have been the drunken boasts of an arrogant individual," Hollingsworth wrote in the memos, "[i]t is known that Gerrard makes substantial cash payments and he also has expensive tastes. He owns several horses and recently bought an Aston Martin for £120,000 in cash."

The anonymous claims made about Gerrard in 2012 were never substantiated. And nearly a decade later, ENRC was still after the attorney. Gerrard, who rejected any suggestion of wrongdoing in his representation of ENRC, charged in a 2019 lawsuit that the company hired Diligence, the firm that in the mid-2000s had conducted the James Bond–style caper to trick a KPMG auditor to spy on him. Among other tactics, Gerrard said, Diligence operatives planted a tiny, motion-sensitive video camera in a tree on his property that filmed every car entering it, recording its license plates in the process.

Gerrard also alleged that Diligence spies had tailed his wife and him when they left England for a Caribbean vacation and had been aboard the same flight, seated a few rows behind them. The Diligence operation came to light because the firm's operatives screwed up. The Caribbean island where Gerrard and his wife were staying was private and the operatives, in order to convince authorities to allow them on it, claimed to be relatives of the couple. The trouble was they referred to Gerrard and his wife by their first names, rather than by their middle names, which was how they were known. Authorities inspected the bags of one Diligence spy and found it stuffed with electronic surveillance equipment, including a camera outfitted for nighttime use. Gerrard claimed in his lawsuit that ENRC had also hired Black Cube to spy on him. In court filings, ENRC said that all the investigative activities conducted on its behalf by Diligence and Black Cube were legal and legitimate.

AROUND THE SAME TIME that Neil Gerrard filed his lawsuit, a newsletter that covers the private investigations industry, *Intelligence Online*, published an article about Mark Hollingsworth. The publication provides a venue for firms to publicize new clients and for operatives to announce new ventures. The article about Hollingsworth was different. It reported that someone had hacked into Hollingsworth's email account and that investigative firms in London for whom he had worked as a contractor had received a note containing a dump of his emails and an ominous-sounding message.

The message was addressed to British authorities and the brief note accompanying it read as though it came from a bomb thrower bent on exposing the private spying industry. It warned that "black," or corrupt, lawyers and judges in England were using hired operatives to pervert the justice system.

The email's subject line said "Episode 1: double agent," and was sent from an account named sam@thetruthprovider.com. The accompanying text read:

Dear UK authorities,

Your work is destroyed by double agents working for corrupt investigations agencies;

Alaco
Control risk
Dangate
Diligence
Fusion GPS
Good Governance Group
GPW
Greyhawk
Insideco
Kroll
K2
Martello risk
Orbis

These bandits are hired by black lawyers and Judges. All of these persons need to be held to justice. Good bye and good luck with arrests I will help more in the future when I see your actions.

I had first met Mark Hollingsworth in London in early 2019 to talk about this book. He was aware that Glenn Simpson and

Christopher Steele had refused to speak with me but he said that as someone who had spent his life as a journalist, he felt an obligation to try to help. At the time of our meeting, he was still shaken by the death of his wife, who had succumbed to cancer a year earlier. He and Glenn Simpson had parted ways and Hollingsworth said he hadn't heard from Simpson after his wife's death.

Later on, after *Intelligence Online* published its piece about the hack of Hollingsworth's email, he insisted to me that it wasn't a big deal. Financial con men had hacked into his email before and tried to scam his friends into sending them money, he said, and the current episode seemed like a repeat performance. "This is amateurville," he remarked. He also didn't think the hack was in any way connected to the ENRC case. "People are really bored of the whole Kazakh story," he said.

HOWEVER, THE MATERIAL IN the dump of Mark Hollingsworth's email suggested that someone was very interested in it. The electronic file attached to the "Episode 1: double agent" email was huge, containing 2,225 pages of material. Much of it was junk—policy reports, lawsuits, and other extraneous material that Hollingsworth had sent to people as email attachments. But there were also dozens of emails Hollingsworth had exchanged over the years with Glenn Simpson, Christopher Steele, Alex Yearsley, Charlie Carr, Rinat Akhmetshin, and other private operatives and reporters. Many of them had to do with ENRC, the Kazakh Trio, or their related company, International Mineral Resources.

THE EMAIL DUMP ALSO contained messages between Hollingsworth and a private operative in Washington, D.C., named Phillip Van Niekerk. Those emails indicated that the three oligarchs comprising the Trio had a falling out in 2017 and that Van Niekerk was hired by a U.S. law firm retained by one of the oligarchs to investigate his partners. Van Niekerk then hired Hollingsworth to work with him on the case.

Hollingsworth wrote an article for a British newspaper, the *London Evening Standard*, reporting that the Serious Fraud Office had interviewed one member of the Trio about his associates. "I have also arranged for this to be published in next week's 'Intelligence Online,'" Hollingsworth wrote Van Niekerk. "And I have been approached by a Bloomberg reporter, about it."

As part of his assignment, Van Niekerk was supposed to identify private operatives working against his client's interests. Rinat Akhmetshin, the Russian-born consultant was one of those targets and Van Niekerk later claimed that Hollingsworth suggested they approach Akhmetshin to see if he was willing to flip sides and become a paid informant. Hollingsworth disputed that account, but soon after his article about the Trio appeared in the *Evening Standard* he reached out to Akhmetshin. "As you know, I am always happy to share information with you," Hollingsworth wrote. "I was just wondering if you have heard anything about the Kazakh Trio in the past few months or have any new information or documents." Akhmetshin didn't respond. "Mark is a fucking rat," he said when shown the email.

Right around the same time Hollingsworth emailed Akhmetshin, he sat down at a computer and composed an extensive memo describing his various assignments from Van Niekerk. The memo, however, wasn't intended for Van Niekerk. Instead he referred in it to the operative, who was from South Africa, by the initials "SA."

Hollingsworth wrote that he had concluded that the source of information about a meeting between a Trio member and the Serious Fraud Office was the American lawyer who hired "SA."

The law firm and "SA," he wrote, were working on behalf of one member of the Trio, adding that "SA" had repeatedly pressed him to plant negative news articles about the oligarch's two partners. He also described how "SA" had strategized with him on how to turn Rinat Akhmetshin into an informant. The "client is prepared to pay a large consultancy fee if Rinat co-operates," Hollingsworth wrote.

PHILLIP VAN NIEKERK NEVER sensed Hollingsworth's betrayal and, for two years, the journalist/private operative acted as if nothing between them had changed. And in 2018, Hollingsworth introduced Van Niekerk to another hired spy for whom he worked as a contractor—Christopher Steele. By then, Steele had gained fame for his role in the dossier and was insisting to journalists that he was certain that his Trump/Russia reports were "70 to 90 percent" accurate. His business partner, Christopher Burrows, who was more personable than Steele, joked with one reporter that he put the figure at 50 percent. "I just don't know which 50 percent," he'd add dryly.

When Van Niekerk met Steele, the lawsuit filed in London against Orbis Business Intelligence by the oligarchs who controlled Alfa Bank was moving forward and Steele was looking for ammunition. Hollingsworth had told him that Van Niekerk knew a former Russian intelligence agent who had defected to the United States, where he now worked as a private operative. That ex-Russian spy, Yuri Shvets, had once investigated Alfa Bank and Van Niekerk said Steele asked him to put him in touch with Shvets. Van Niekerk never learned if the two men ever spoke but it's unlikely Steele would have gotten a warm reception from the ex-Russian agent. Not long after BuzzFeed posted Steele's memos, Shvets described the reports in one interview as having "zero" credibility.

Right around the same time as the dump of Mark Hollingsworth's emails something else happened: ENRC sued him. The
company charged that Hollingsworth had conspired with Robert
Trevelyan, the computer expert known as "Magic," to shop confidential records Trevelyan had taken from the company. The lawsuit
contained snippets of emails exchanged between Hollingsworth
and Glenn Simpson, Philip Van Niekerk, as well as other operatives.

In addition, ENRC filed separate cases in U.S. courts against
Simpson and Van Niekerk, demanding that they provide the company with information about their interactions with Hollingsworth. And tucked in those filings, the company made a big reveal:
Mark Hollingsworth and Robert Trevelyan had both been secretly
working for ENRC as informants. The company didn't provide
details about when ENRC recruited the men, though the filing
stated they were paid monthly consulting fees. Trevelyan was still
in the company's good graces because he hadn't been sued. As for
Hollingsworth, ENRC claimed that he had "breached" his agreement with the company, though it didn't elaborate.

Whatever occurred, someone had decided to throw Hollingsworth under the bus and end his dual career as a journalist and
private spy. Phillip Van Niekerk only realized that Hollingsworth
had flipped on him when he got served with ENRC's filing in
2019. The company claimed in its filing seeking documents from
Van Niekerk that Hollingsworth had fingered him as his source
for the *Evening Standard* article about the Trio, something that
Van Niekerk denied. He was furious and called Hollingsworth to
confront him. He said that Hollingsworth claimed that operatives
working for ENRC had lured him into a trap. "He said he had
gotten in over his head," Van Niekerk said.

One of those private spies, emails indicated, was a Moscow-
based operative named Dmitry Vozianov. "There are not many
people in the corporate intelligence industry who inspire as much

awe," as Vozianov, an article in *Intelligence Online* stated, describing the operative as a bare-knuckled brawler who largely worked for oligarchs. (*"Be careful if you write him,"* the publication said it was warned while reporting that article.)

In emails, Vozianov urged Hollingsworth, among other things, to generate articles attacking enemies of ENRC including the lawyer, Neil Gerrard. "We need very much strong language and aggressive articles," Vozianov wrote in one 2018 message. "Including 'SFO conspiracy with Gerrard among other things. . . .'"

In hindsight, Phillip Van Nickerk suspected that Hollingsworth had probably started working as an informant for ENRC back in late 2017, the time when the journalist/operative wrote his "SA" memo. When he finally saw that memo, he maintained that little in it was true and that it read as if Hollingsworth had written it to make his new paymasters happy. Van Niekerk knew his profession could be treacherous but the depths of Hollingsworth's deceptions went way beyond anything he had experienced. "I'm blown away by the duplicity," he said.

In defending himself against ENRC's lawsuit, Hollingsworth said in court papers that he hadn't done anything wrong and that the company was misrepresenting his actions and words. He insisted he had obtained the company's documents in his capacity as a journalist, didn't know how "Magic" had acquired them and hadn't profited from sharing the records with others. He told me in an email that he had been forced to work as a private operative in order to take care of his wife during her bouts with cancer. It was hard not to be sympathetic to his situation. But some people who had gotten burned by Hollingsworth said that didn't excuse his behavior. The Global Witness investigator, Daniel Balint-Kurti, would discover from one of Hollingsworth's hacked emails that the journalist/private spy had passed on confidential information he had shared with him to another operative-for-hire. "That was pretty low," Balint-Kurti said.

CHAPTER 15

SHINY OBJECT

NEW YORK CITY, 2019

In 2019, two books about the Trump/Russia controversy and the dossier were published and both briefly became *New York Times* bestsellers. As was the case with many books that came out during the Trump years, they occupied polar opposite ends of the political spectrum.

For people who loved Trump, the book for them was *The Plot Against the President*. Its hero was Congressman Devin Nunes, the California Republican viewed by liberals as a Trump flunky. But in the book's telling, Nunes is a truth-seeking hero who bravely uncovers a "deep state plot to bring down a new President." Written by Lee Smith, a former journalist who now worked at a conservative think tank, *The Plot Against the President* asked some valid questions, like how Christopher Steele managed in just a few weeks to unearth a complex conspiracy between the Trump campaign and the Kremlin. But Smith was selling a far-fetched conspiracy of his own. In his dark alliance, Steele wasn't the author of the dossier but a sock puppet for rogue government officials and the bearer of their handiwork. "What I'm saying is that we have the order wrong," Smith said in one interview. "It starts with people who know how to write something, how to write a document to get [national security] warrants. It's not like it comes out of the blue from Christopher Steele who is finding all this marvelous stuff and has to go to the FBI. It is the other way around."

For people who despised Trump, Glenn Simpson and Peter Fritsch's *Crime in Progress* was their cup of tea. The book was a hagiographic memoir that cast Simpson, Fritsch, and Christopher Steele as modern-day Musketeers out to save democracy. "I've read kind of all the books on this subject," Rachel Maddow proclaimed, "and this is the one you want to read." And the book found plenty of readers. In publicizing it, Simpson and Fritsch presented a carefully-crafted version of events. One theme that Steele struck in his memos—that the Kremlin was trying to interfere with the U.S. election and supporting Trump—was rock solid. U.S. intelligence and law enforcement agencies had come to that same conclusion and there was plenty of evidence of Russian meddling, from the hack by Kremlin operatives of Democratic campaign computers to an assault of fake bots launched by a Russian troll farm.

But the real juice that set Steele's dossier apart was its central charge that Trump campaign staffers and Kremlin operatives had colluded in order to help Trump and that Vladimir Putin had blackmail material on him. That was the story that Fusion GPS had sold to journalists before and after Election Day. It was also at the heart of the book that Simpson and Fritsch wrote.

But by the time of the book's appearance, the air was going out of the collusion balloon. In the spring of 2019, Robert Mueller III released his long-awaited report on Russian efforts during the presidential election. He found that Moscow did try to influence voters to favor Trump but he didn't find evidence of criminal collusion and his report was virtually silent about the dossier.

Mueller's report made it clear that one of the most consequential charges contained in the dossier—the one about Michael Cohen's supposed trip to Prague—was dead wrong.

Cohen lied about a lot of things and he would eventually plead guilty to lying to Congress about secret efforts during the 2016 campaign to strike a deal to build a Trump Tower in Moscow. That initiative provided a simpler way than a tangled conspiracy theory to understand Trump's subservient behavior toward Vladimir Putin. Trump didn't expect to become president so he was sucking up to Putin for another reason: To make money.

THE DAY AFTER THE Mueller report was disclosed, *The New York Times* reported that the inquiry's findings about the dossier—or more specifically, its lack of them—would intensify scrutiny of Steele, his reports, and his sources of information. "Some of the most sensational claims in the dossier appeared to be false, and others were impossible to prove," the *Times* reported. An article about the Mueller report in *The Wall Street Journal* sounded a similar theme. The report, it stated, "all but dismissed many key claims of the dossier."

A lawyer for Fusion GPS issued a response that focused on the dossier's general theme of Russian election meddling rather than its specific claims. "Russia was conducting a covert operation to elect Donald Trump, and . . . the aim of the Russian operation was to sow discord and disunity in the U.S.," the lawyer, Joshua A. Levy, told *The Washington Post*. "To our knowledge, nothing in the Steele memoranda has been disproven."

While publicizing their book, Glenn Simpson and Peter Fritsch struck a similar tone. "To look at an intelligence document like that and fault it for not having total verisimilitude with the facts is kind of an unfair yardstick," Fritsch told *New York* magazine. "The foundational claim of the dossier is almost impossible to dispute, sitting here at the end of 2019."

WHILE GLENN SIMPSON AND PETER FRITSCH were promoting *Crime in Progress*, another U.S. government investigation involving the Trump/Russia controversy was wrapping up. The inspector general of the Justice Department, Michael Horowitz, had earlier announced that his office had started a probe to examine two questions. The first was whether FBI officials had a sufficient basis in 2016 to commence their investigation of Trump staffers. The second was whether the bureau, once it began that inquiry, had followed appropriate procedures.

In his report, which was issued in December 2019, Horowitz upheld the FBI's decision to start its counterintelligence inquiry. But he sharply criticized bureau officials for the methods they had used to obtain warrants to surveil Carter Page, the former Trump campaign advisor, including submitting material from the dossier to a judge in order to get warrants.

Several months before the Horowitz report was released, staffers working for his office had interviewed Christopher Steele. And a few days afterward, Natasha Bertrand, the *Politico* reporter, wrote that unnamed sources had told her those talks had gone swimmingly. "The interview was contentious at first, the sources added, but investigators ultimately found Steele's testimony credible and even surprising," wrote Bertrand. "The extensive interview with Steele, and the investigators' sense that he offered new and important information, may dampen expectations among the president's allies who've claimed that Steele's sensational dossier was used improperly by the bureau to 'spy' on the campaign."

THAT MAY HAVE BEEN the story Fusion GPS wanted to get out. But the findings of the Horowitz report were devastating for Steele and the dossier. The report disclosed, among other things, that Justice Department officials had gone to London in November 2016, not long after severing ties with Steele, to find out more about the ex-spy from his former colleagues at MI6. At that time, the FBI's inquiry into the Trump campaign was still secret but bureau officials wanted to assess Steele's reliability in order to better gauge how seriously to take the allegations in his memos. The "bottom line is that we had concerns about the reporting the day we got it," one top FBI official told Horowitz's investigators in explaining the London trip. Some "of it was so sensational, that we just, we did not take it at face value."

Steele's former colleagues vouched for his honesty and integrity. Still, not all their comments were positive. Some MI6 officials noted he had a tendency to become obsessed and go down rabbit holes chasing targets of questionable value, traits that he and Glenn Simpson seemed to share. "To the extent there were negatives," one FBI official said afterward, "[i]t was that he was the type of person who would sometimes follow the shiny object, without, perhaps, a deep set of judgement about the risk that may or may not be there."

In addition, the Horowitz report revealed that the FBI, shortly after BuzzFeed posted Steele's reports, had located his "collector" and interviewed him. And Steele's key informant did exactly what an operative—or a journalist—prays a critical source never does: he offered a different story. The ex-spy's collector, who wasn't named in Horowitz's report, told FBI agents in early 2017 that he hadn't seen Steele's memos until BuzzFeed published them and was shocked when he did. He claimed that the ex-spy had "misstated" the nature of the information he had provided to him or "exaggerated" its reliability, saying the material was often "just talk," "word of mouth and hearsay," or "conversations that [he] had with friends over beers." After the 2016 election, the man added,

Steele had asked him to seek verification for the dossier but he had found "zero."

Steele was outraged by his depiction in the Horowitz report. Orbis Business Intelligence put out a statement saying the report contained "numerous inaccuracies and misleading statements" and minor changes were made to it. But those fixes were a small victory because the Horowitz report also suggested that Kremlin agents might have locked on to Steele's sources when they started inquiring about Donald Trump—the scenario Scott Shane, the *Times* reporter, had painted at the Spy Museum. Russian intelligence operatives were likely monitoring the activities of Orbis Business Intelligence and Steele's allegations about Michael Cohen's trip to Prague could have been a plant, the report said.

Steele was adamant that his memos hadn't been contaminated by Russian disinformation. But if so, the other reason for its inaccuracies wasn't particularly appealing. If the material wasn't disinformation, there was only one alternative. It was shit information.

TWO PEOPLE HAD REFUSED to be interviewed by investigators working on the Horowitz report—Glenn Simpson and Jonathan Winer, the lawyer who had arranged Steele's State Department meeting. Simpson had decided to tell his story in a format he could control—the book that he and his partner, Peter Fritsch, wrote. As for Winer, he had gotten himself tangled up in a mess.

A month before the 2016 election, Sidney Blumenthal, a longtime Clinton family operative, gave two memos to Winer about Donald Trump's supposed sexual antics in Russia. The memos were written by Cody Shearer, a self-styled private investigator who some journalists viewed as a loose cannon. Shearer stated in one memo that a supposed FSB source had told him that Donald Trump was secretly filmed twice by the spy agency having sex. "He said he believes a copy of the sex videos is in Bulgaria, Israel and FSB political unit vaults in Moscow," Shearer wrote. He said his source claimed he saw the video and described Trump's behavior as "very dark."

Instead of throwing out Shearer's memos, Winer gave them to Christopher Steele, who then passed them on to the FBI. After Winer's actions became public, he wrote in an op-ed piece in *The Washington Post* that he never thought that Shearer's memos would "be shared with anyone in the U.S. government." Winer also told Congressional investigators that Steele asked him after BuzzFeed published the dossier to return all the Orbis reports he had sent to the State Department or make sure they were destroyed so the reports didn't become public and identify his sources. Winer didn't purge the records from the State Department's system but he acknowledged in testimony that he had deleted all his correspondence with Steele from "his personal devices."

BY 2019, THE DOSSIER had also become a hot potato within the news media. Conservative publications such as *The Daily Caller* and *The Federalist* had long written skeptically about Steele's claims but now a few journalists who once embraced his memos were backing away. Prime among them was Michael Isikoff, the *Yahoo! News* reporter who had written the first article about the dossier. In 2018, the decades-long friendship between Isikoff and Glenn Simpson abruptly ended. In *Russian Roulette*, the book Isikoff wrote with David Corn, he had paraphrased Simpson as describing Steele's collector as a "Russian émigré living in the West who travelled frequently to Moscow." Simpson was furious when his comments appeared in the book, believing that he told his friend the information was "off the record." Simpson stopped speaking to Isikoff and Peter Fritsch sent him expletive-filled emails.

Isikoff soon took another step that couldn't have thrilled

Simpson or Fritsch. In late 2018, he began to publicly question the dossier. "When you actually get into the details of the Steele dossier, the specific allegations, we have not seen evidence to support them," he said in one interview. "In fact, there is good grounds to think that some of the more sensational allegations will never be proven and are likely false." After the appearance of the Mueller report, Isikoff added that journalists, including himself, should have scrutinized Steele's claims more carefully. "I think it's fair to say that all of us should have approached this, in retrospect with more skepticism, particularly when we didn't know where it was coming from," he told another interviewer. "It's been surprising to me the degree to which some people have wanted to maintain that the dossier was checking out when, as far as I can tell, it hasn't."

Those comments put him at loggerheads with some of his former fellow travelers. Isikoff was one of Rachel Maddow's guests during the special episode in 2017 of her MSNBC show devoted entirely to the dossier. Two years later, in 2019, he invited Maddow, who was then promoting a new book she had written, to be a guest on his own podcast.

During their interview, he asked her whether she now believed she had oversold Steele's reports to viewers. "Do you accept that there are times that you overstated what the evidence was and you made claims and suggestions that Trump was totally in Vladimir Putin's pocket?" he said.

An acrimonious exchange ensued. "What have I claimed that's been disproven?" Maddow asked.

"Well, you gave a lot of credence to the Steele dossier," he said.

"I have?" she replied.

"Well, you've talked about it quite a bit," he went on.

Things got more heated. "You are trying to litigate the Steele dossier through me as if I am the embodiment of the Steele dossier, which I think is creepy, and I think it's unwarranted," Maddow said. "It's not like I've been making the case for the accuracy of the Steele dossier and that's been the basis of my Russia reporting."

FOR OTHER JOURNALISTS, THE report by Michael Horowitz, the inspector general, proved the tipping point. The media critic for *The Washington Post*, Erik Wemple, said that the report's rejection of the dossier's specific claims led him to write a series of columns about the media's infatuation with it. And as part of that series, he reached out to journalists and talk show commentators to see if they had any regrets about how they depicted Steele and his reports.

Howard Fineman, a longtime political journalist in Washington, was one of the few people who voiced regrets. "The Horowitz Report is a devastating critique of the Steele Dossier," Fineman said. "It's clear that the whole emergence and effect of the dossier should be a warning to us all: bad oppo + zealous feds = bad journalism, erosion of the rule of law and threats to civil liberties."

But most of the journalists and commentators contacted by Wemple either blew him off or defended their reporting. Jane Mayer of *The New Yorker* told him that Steele had become a convenient "political pinata" for Trump supporters, adding that her article had "carefully quoted naysayers." Wemple devoted an entire column to what he described as Natasha Bertrand's overheated and inaccurate coverage of the dossier, arguing that she had turned it into an opportunity for career advancement. "With winks and nods from MSNBC hosts, Bertrand heaped credibility on the dossier . . . in repeated television appearances," Wemple wrote. "Along the way, she bootstrapped her punditry into a contributor's role on MSNBC." Bertrand and her editor at *Politico* sent him statements defending her reporting and, after Wemple's column was published, Bertrand sent him an email. "This is really low even for you Erik," she wrote. "I got my msnbc [*sic*] gig through hard work reporting on a range of issues related to Mueller, not through 'bootstrapping' the dossier. I hope it was worth the 3 retweets."

ALONG WITH RIGHT-WING PUBLICATIONS, some left-wing journalists also had long been skeptical of the dossier. One of them, Matt Taibbi, a onetime political writer for *Rolling Stone* magazine, cast the media's handling of Steele's reports as a replay of a previous press disaster—the reporting in the early 2000s in *The New York Times* and elsewhere that had backed false claims by the Bush administration that Saddam Hussein possessed weapons of mass destruction, or "WMDs."

"The WMD affair showed what happens when we don't require sources to show us evidence, when we let political actors use the press to 'confirm' their own assertions, when we report on the journey of rumors instead of the rumors themselves," Taibbi wrote about the media's treatment of the dossier. "When we let stuff like this go, the public sees us as fools, at which point it doesn't matter whether what we write is for or against any politician, because nobody believes us anyway. Is this really the industry standard we're gunning for? Are we never going to own up to this one?"

The short answer to that question was no, though politics played only a small part. Media organizations didn't conduct internal postmortems or public reconstructions about how they handle the dossier story for another reason—it would have required them to disclose the toxic relationship that had developed between journalists and private spies. Those ties weren't new; ever since the days of Jules Kroll, private operatives have thrown out tips to eager reporters who, in turn, have concealed the involvement of a paid operative in a story. Glenn Simpson and Peter Fritsch had simply been better at playing reporters and no journalist was about to admit that.

To make matters worse, some outlets didn't even bother to correct errors in dossier-related stories. Few news organizations had become as invested in the Michael Cohen in Prague story as the McClatchy chain of newspapers, whose publications included the *Miami Herald*, the *Sacramento Bee*, and the *Charlotte Observer*. In 2018, two McClatchy journalists reported that Robert Mueller's staff had confirmed parts of the dossier by finding "that Cohen entered the Czech Republic through Germany, apparently during August or early September of 2016." The same reporters doubled down with a follow-up article in which they reported that a foreign intelligence service had found evidence that cellphone towers in Prague had registered pings from Cohen's phone.

But after the Mueller report, the newspaper chain, instead of acknowledging its mistakes, issued a begrudging editor's note. It stated: "Robert S. Mueller III's report to the attorney general states that Mr. Cohen was not in Prague. It is silent on whether the investigation received evidence that Mr. Cohen's phone pinged in or near Prague, as McClatchy reported."

The New York Times also found its reporting challenged. In February 2017, the paper published an article with the headline, "Trump Campaign Aides Had Repeated Contacts with Russian Intelligence." A month later, James Comey, the FBI head, disputed the accuracy of the article during the course of his congressional testimony but the *Times* stood its ground.

For its part, *The Guardian* never addressed what some journalists believed was a major miscue by its star reporter on the Trump/Russia beat, Luke Harding. In 2018, he and two colleagues published a scoop that looked for a moment like a game changer. They reported that Paul Manafort, Trump's campaign manager, had held a secret meeting in 2016 with Julian Assange, the founder of WikiLeaks, the group that would publicize the Democratic Party emails obtained by Russian hackers.

The story was all the more sensational because it involved a Houdini-like feat. At the time of the supposed meeting, Assange was living inside the Ecuadorian embassy in London, where he had taken refuge to avoid arrest. Both the embassy and Assange's room were monitored by security cameras, which meant that Manafort's arrival should have been captured on tape. Harding and his colleagues wrote that the meeting "is likely to come under scrutiny and could interest Robert Mueller." Harding, who had a flair for self-promotion, tweeted out, "Manafort visited the embassy in 2013, 2015, and, in spring of 2016, according to multiple sources and docs [documents]. Last visit took place around the time he joined Donald #Trump's campaign."

The trouble was that no other news organization was able to confirm Harding's reporting because the visits apparently never took place. And when reporters asked editors at *The Guardian* to explain how the article had gotten into the paper, they got stiff-armed. Years later, the newspaper said it was still standing by its reporting without revealing why it was doing so.

THE CREDIBILITY OF CHRISTOPHER Steele's reports soon suffered another blow and this time the criticism came from Fiona Hill, the former State Department official who emerged in 2019 as the star witness of the impeachment proceedings against Donald Trump. Hill, a Russia expert, gave a deposition to congressional staffers in which she said she had known Steele for years because they had been government counterparts on Russian intelligence matters. After he became a private operative, she added, he regularly contacted her looking for work. "He was constantly trying to drum up business," she remarked.

Her assessment of the dossier was pretty dim. Hill, who was lauded as a straight shooter for her frankness, thought that Steele had gone down a "rabbit hole" and that the Kremlin may have fed his sources disinformation. She was also taken aback by Steele's decision to involve himself in a U.S. political campaign. "He's a former foreign spy," said Hill, who grew up in England. "But nonetheless, a foreign national. I don't believe that it was appropriate for him to have been hired to do this."

CHAPTER 16

DINNER WITH NATALIA

MOSCOW, 2019

Natalia Veselnitskaya looked trimmer than she had in television interviews after the news broke in 2017 about her participation in the Trump Tower meeting. She hadn't been in New York for some time and her daily routine of eating at Nello and Ess-a-Bagel had ended. "I've lost twenty kilos," Veselnitskaya said.

We were having dinner in Moscow's financial district at a fish and sushi restaurant called Bamboo. Rinat Akhmetshin, who arranged our meeting, was there, as was a translator who worked for Veselnitskaya. I had gone to Moscow because Veselnitskaya wasn't coming back to New York. In 2019, the Justice Department had indicted her for obstruction of justice and she faced arrest if she set foot again on U.S. soil. The charge had nothing to do with the Trump Tower meeting or the 2016 presidential election. Instead, she was accused of concealing information in a court filing connected to the case against Prevezon, the real estate firm suspected of laundering money stolen from Bill Browder's investment fund.

In 2017, the Justice Department and Prevezon settled charges against the company and, as part of the deal, Prevezon, which denied any wrongdoing, agreed to pay a $5.9 million fine. That was a lot less than federal prosecutors had initially hoped to get, and from Veselnitskaya's perspective the case's ending looked like the beginning for her of a new international legal career.

Instead, she had ended up under indictment. She could blame hackers and Microsoft for that. After the Prevezon case ended, someone hacked into her computer and grabbed her files, which eventually ended up at the Justice Department. In those records, officials said they found evidence that Veselnitskaya had secretly collaborated with Russian prosecutors to prepare a court filing made in the Prevezon case in which Kremlin authorities claimed that they had investigated Prevezon and cleared it of any wrongdoing. The giveaway was contained in a feature of Microsoft Word called "track changes." The tool records changes made in a document as it goes through different drafts—deletions, additions, etc.—and federal officials saw the changes Veselnitskaya had made in the Russian government filing. She then was indicted for allegedly failing to disclose that she had coordinated her work as a defense lawyer with Kremlin prosecutors.

She didn't appear too upset by her predicament. She said that U.S. prosecutors had come after her because she humiliated them by forcing a settlement of the Prevezon case for far less than they had wanted. The allegation was bogus, she added, and she expected it would get thrown out.

Over dinner, we talked about a number of subjects. Veselnitskaya said she still had warm feelings for Glenn Simpson and considered him a talented investigator who had dug up critical information for Prevezon. As for Christopher Steele, she had nothing but contempt for him and considered the dossier a farce.

"Do you really believe that an ex-employee of MI6 who has never been to the Russian Federation [for decades], who has no connections to the Russian Federation, who has no friends in the Russian Federation, neither in the government structure or the legal structure nor even in the prosecutor's offices, do you really believe that he was really capable of running an investigation that took him only two months about a five-year-long connection between Trump and Putin," she asked, before pausing to allow her translator to catch up. "If you take this fake stuff for real stuff then you have to be just brave enough to believe, to completely dismiss all your special services, all your intelligence staff, all your FBI, CIA, the national security agency, to dismiss all of them because all those people who most obviously have their people in the Soviet and all their structures could never find out what that talented person managed to find out without ever leaving his room."

Her assessment struck me as a bit harsh but I got her point. She made another observation. If Steele was such a hotshot, she added, why hadn't he or his well-wired sources caught wind of something going on right under their noses—her meeting at Trump Tower. "The so-called Steele dossier or the Trump dossier, it was written during the period when I was meeting" with Donald Trump Jr., she said.

The lawyer described the period following the disclosure of her role in the Trump Tower meeting as a difficult one because she had been portrayed in the media as a Russian spy. But more recently, she added, Americans were taking a far more favorable view of her. She knew that from responses to her new website. It was called "Vensa," which means spring in Russian, and the site was largely devoted to her continuing campaign against Bill Browder.

"I have a lot of followers, a lot of readers and visitors and I can judge from the reactions of common American people who in the very beginning used to call me a witch," she explained. "But then, gradually step by step, I started posting evidence, documents, transcripts of depositions. I could see that common people who I never met, just regular Americans who can read. More and more they were starting to ask themselves the same question, 'How come it was possible for someone to put us in a position where, figuratively speaking, the tail was wagging the dog?'"

I didn't mention that I had checked before coming to Moscow. It was pretty unimpressive. There were short video clips posted of young people pretending to be news announcers, reading stories about the depths of Browder's lies. Crime-show-style charts connected people with ties to the Magnitsky case and the articles on the site had just a few hundred hits.

"What makes Americans strongest and weakest is that they are gullible," Veselnitskaya said. "The nation as a whole, you easily buy into something and you buy into the most monstrous lies. This is your greatest weakness and your greatest strength is when you understand that you have been used as a nation, that you realize that your weakness has been used against you and you simply annihilate those who did so."

After dinner, I wished Veselnitskaya the best and offered to take her to dinner at Nello once her legal problems were over. That wasn't likely to happen soon. A bipartisan Senate report concluded in 2020 that her ties to the Kremlin and Russian intelligence services were "far more extensive and concerning" than previously known.

RINAT AKHMETSHIN AND I left Bamboo and caught a cab. The disclosure of the Trump Tower meeting had turned his career upside down. U.S. prosecutors had questioned him during their examination of possible collusion between the Trump campaign and Moscow. They had cleared him of any wrongdoing but by then all the publicity had proved radioactive.

He was still getting some work. A former Kazakh official paid him $60,000 to try to convince U.S. lawyers that he was the victim of a political attack. But Akhmetshin said that some former clients didn't want to deal with him anymore and old contacts weren't returning his calls. He blamed the news media, saying reporters had denigrated him and implied that he was a Russian spy. "They treated me like I was some kind of piece of shit who had just gotten off the boat," he said.

Akhmetshin, like Veselnitskaya, was eager to avenge himself against Bill Browder. He had sued the investor in a U.S. court, claiming that Browder had libeled him by describing him as a Russian intelligence agent, but the case was dismissed for jurisdictional reasons, a decision Akhmetshin had appealed. Meanwhile, hoping to reinvent himself, he had set up a nonprofit foundation modeled after the Anti-Defamation League and whose stated mission was to combat prejudice against Americans of Russian descent.

Akhmetshin had come to Moscow to try to drum up business, though he confessed he wasn't getting many takers. The falloff in business meant adjusting to the loss of the trappings that had been a part of his life as a private operative—the lavish meals at gourmet restaurants, the expensive and rare wines, the business-class or better air travel. To save money on his flight between Washington, D.C., and Moscow, he had traveled on TAP Air Portugal, the discount carrier that requires flyers to changes planes in Lisbon. The same Senate intelligence panel report that described Natalia Veselnitskaya as having ties to Russian intelligence also suggested that Akhmetshin had them too, a suggestion he rejected. "Just because you work in a zoo and get to know lots of monkeys, doesn't mean you work for them," he said.

BY THE START OF 2020 and a new decade, lawsuits involving corporate intelligence firms and private spies were flying fast and furious. Claims filed by ENRC against Mark Hollingsworth and others were inching ahead, as were the claims filed by the lawyer Neil Gerrard against ENRC and Diligence for allegedly spying on him. Rose McGowan, one of Harvey Weinstein's accusers, was suing Black Cube, David Boies's law firm, and others, alleging they had conspired to dupe her. All those sued denied wrongdoing.

The first case to reach trial was the lawsuit brought in London by the founders of Alfa Bank against Orbis Business Intelligence. It involved a dossier memo written by Christopher Steele in September 2016 that was titled "RUSSIA/US PRESIDENTIAL ELECTION: KREMLIN-ALPHA GROUP COOPERATION." The memo claimed that the bank's principals and Vladimir Putin had long been in a back-scratching relationship. "Significant favours continued to be done in both directions, primarily political ones for PUTIN and business/legal ones for Alpha." It also charged that the bank's founders had used an employee in the 1990s as a bagman to make payoffs to Vladimir Putin, when he was mayor of St. Petersburg.

The "pinging" story that Glenn Simpson and Steele had shopped to reporters and U.S. officials wasn't at issue in the lawsuit because it wasn't part of any dossier memos. Still, as litigation against Fusion GPS and Orbis Business Intelligence was under way, allies of Simpson and Steele were continuing to sell the story and the point person on that effort was Dan Jones, the former congressional aide. By 2018, the Democracy Integrity Project, the "dark money" group headed by Jones, had attracted scrutiny from *The Daily Caller* and other right-wing publications. And soon another "dark money" group called Advance Democracy, that Jones also headed, started. Tax filings made by Advance Democracy showed that donations to that group were also passed on to Fusion GPS and an entity affiliated with Christopher Steele.

Jones commissioned outside experts to study the "pinging" incident and he met with *New York Times* reporters to urge them to reexamine it. The *Times* took a pass but he had better luck at *The New Yorker* with a former *Times* reporter, Dexter Filkins, who now worked there. Filkins, a onetime war correspondent, gave the complicated tale of suspected cyber intrigue his best shot. In a lengthy article in 2018, he argued that the explanations put forward by Alfa Bank and some researchers that the "pinging" was random didn't make sense. But while implying something was seriously amiss, Filkins was unable to say what it was. Not long after the article was published, Robert Mueller remarked during congressional testimony that he didn't believe there was anything behind the "pinging" story. "My belief at this point is that it is not true," he said.

TO SUE ORBIS BUSINESS Intelligence in London, the founders of Alfa Bank took advantage of the same British privacy laws that the anti-asbestos activists had used to bring their claim against Rob Moore and K2 Intelligence. Over the past decade, officials in Europe had taken a tougher stance than their U.S. counterparts toward businesses that gather and sell user information. And the result was laws in England and elsewhere requiring Facebook, Google, and other big data companies to notify people about information gathering.

Corporate investigative firms in England were required to go through an exercise prior to starting an inquiry and determine if it could be done without intrusive measures. The law had plenty of exceptions and loopholes—private spies didn't have to notify their targets, for instance, when they were gathering material for use in litigation or employee disputes. Still, the new rules provided people unfairly targeted by operatives-for-hire with the opportunity to seek recompense.

As the Alfa Bank–related lawsuit was starting in March 2020, another development was unfolding—a deadly novel form of coronavirus was spreading around the world. In response to the growing pandemic, parts of the United States and Europe had shut down but Britain was late to do so and the case against Orbis Business Intelligence proceeded. When he took the witness stand, Christopher Steele quickly found himself under attack. The bank's owners denied making payments to Vladimir Putin and their lawyer pressed Steele to explain apparent inaccuracies in his Alfa Bank memo as well as the criticism of his work leveled by the Horowitz report.

In his testimony, Steele appeared to display a consistent attitude—he was right and other people were wrong. Horowitz's investigators, Steele said, had mischaracterized what he had told them. Notes written by Kathleen Kavalec, the State Department official to whom he had spoken about Alfa Bank, had mischaracterized his comments. Bruce Ohr, his old contact at the Justice Department, had misunderstood what he had said to him.

Steele insisted that all the information in his Alfa Bank memo had come through his trusted collector, who had provided high-quality information for six years to Orbis. His reports were so reliable, Steele said, that the firm kept him on a retainer, paying him from $3,000 to $5,000 a month. He batted away questions from the oligarchs' lawyer about the collector's identity, citing concerns for his safety, and insisted that even disclosing the country where he resided would give Russian intelligence a piece of the "jigsaw" that they could use to identify him.

But the problem for Steele involved factual errors in his Alfa

Bank report. As the oligarchs' lawyer pointed out, the bagman who supposedly delivered payoffs to Putin wasn't employed by Alfa Bank during the period in question. Steele defended the substance of the memo but acknowledged the mistake. "I don't agree that the facts given me by the source are false," he testified. "I agree that one of the points made is false."

Steele never got to complete his testimony in the case. Midway through his second day on the witness stand, he was notified that his wife had been taken to a hospital near their home in Farnham after showing symptoms of COVID-19, the disease caused by the novel coronavirus. He was excused so he could join her and the case was completed without him.

Several months later, the judge overseeing the case issued his ruling. He found that Orbis had not violated the Data Protection Act, the law under which the oligarchs had sued. But when it came to the most serious allegation in Steele's report—the claim that Alfa Bank's principals had bribed Putin—the judge held that Steele had failed to take reasonable steps to ascertain whether the charge was true.

"Mr. Steele knew that his source did not have direct personal knowledge of the underlying facts, but could only be relying on hearsay," the judge stated. "He has failed to explain how that information would or could have come to the sub-source by virtue of his job. The allegation clearly called for closer attention, a more enquiring approach, and more energetic checking." He ordered Orbis to pay modest damages of about $22,600 to each of two Alfa Bank–connected oligarchs mentioned in the memo.

IN THE SUMMER OF 2020, Christopher Steele and Orbis Business Intelligence faced a second trial in London and this time the stakes were higher. At issue was another dossier memo the ex-spy had written just after the 2016 election. He said in it that the Russian owner of several internet service providers knowingly allowed Kremlin operatives to use his firm to launch "botnets" and "transmit viruses, plant bugs [and] steal data" from the Democratic Party.

The businessman, Aleksej Gubarev, denied the assertions and sued Steele and Orbis in England for defamation. The core of Steele's defense rested on his claim that he hadn't discussed any aspects of the dossier with the news media once the 2016 election was over, when Orbis's contract with Fusion GPS ended. He insisted that he never anticipated that his memos would become public and that he had nothing to do with BuzzFeed's publication of them.

"We were no longer in the contract with Fusion," Steele testified. "We had no reason to talk to the media."

A lawyer for Gubarev pointed to text messages that Steele had exchanged in late 2016 and early 2017 with David Kramer, the ex-aide to Senator McCain, arguing that they suggested that Kramer had acted as Steele's intermediary in responding to questions about the dossier from reporters with *The Wall Street Journal*, ABC News, and other outlets. Steele insisted Kramer had acted on his own and disputed earlier sworn statements by the former McCain aide that he had shown the dossier at Steele's behest to both Ken Bensinger, the BuzzFeed reporter, and Carl Bernstein at CNN.

Steele acknowledged that he had started speaking with Bensinger in mid-2016 when Bensinger began gathering material for his book about the FIFA soccer scandal. But he said that apart from a single message that Bensinger had sent him around Christmas referring to Trump and Russia, all their interactions had been about FIFA. He also said that the reporter had told him that a Hollywood studio was eager to make a movie based on his FIFA book and that Orbis could get a piece of the action. "He was writing a book and said he had a film contract potentially, and that this might be a lucrative project to do with my company," Steele testified.

"Are you saying that your messages with Mr. Bensinger were only about FIFA?" Gubarev's lawyer asked him.

"That was a business we were dealing with, yes," Steele responded.

Text messages indicated that Bensinger and Steele had originally planned to meet in late January 2017, apparently to discuss FIFA. But then their meeting was suddenly pushed up to January 3. By that time, Bensinger and other BuzzFeed reporters were trying to confirm the dossier. Still, Steele insisted that Bensinger had primarily come to see him about FIFA and testified that he agreed to the earlier meeting because he didn't want to offend Bensinger and jeopardize the deal the journalist had offered him on a FIFA-related movie. "I was basically trying to keep Mr. Bensinger at arm's length, partly because I was afraid he was trying to investigate whether I had been involved in the Trump/Russia investigation and secondly, because I wanted to preserve the commercial opportunity that he presented to me," Steele testified.

In late 2020, the British judge who heard the case issued a ruling in favor of Steele. He held that even if the contents of Steele's memo about Aleksej Gubarev and his company were defamatory, the businessman and his firm had failed to show financial damages as a result of its disclosure. In addition, the judge ruled that the evidence presented by Steele supported his account of his interactions with the news media.

KEN BENSINGER, THE BUZZFEED reporter, wasn't called to testify as part of the Gubarev lawsuit. But had he done so, questions may have been raised about Christopher Steele's version of events.

Bensinger said Steele knew by early December 2016 that he was interested in the dossier because he contacted him soon after Glenn Simpson told him about it at the Fusion GPS staff retreat. In an initial phone call, Bensinger said, Steele evaded his questions about the dossier. However, the ex-spy was aware, Bensinger said, that their meeting on January 3 was largely about the dossier.

Bensinger added that Steele, during his testimony, also mischaracterized events related to the FIFA film project linked to his book. A Hollywood studio had optioned the book's proposal and Bensinger said that he told people about that deal in order to encourage them to speak with him. But he never suggested to Steele that there was money in it for him. Instead, Bensinger added, it was Steele who called him in late 2016, offering to sell information and suggesting that the film studio might want to hire him as a consultant. "He said he had some intelligence about how Russian organized crime figures were involved in the building of stadiums for the World Cup and he needed a sponsor to develop it further," Bensinger said. The reporter said he told Steele that journalists, unlike his clients in the private spying world, didn't pay for information. Besides, Bensinger would later say, Steele really didn't know that much about FIFA.

CHAPTER 17

THE COLLECTOR

NEW YORK, 2020

Throughout the tangled story of the dossier, its best-kept secret was the identity of Christopher Steele's "collector," his trusted paid informant. Ultimately, every word written by every journalist about the dossier, every opinion expressed about every commentator about Steele's memos, and every declaration by every politician about the ex-MI6 agent's reports tied back to Steele's collector and his abilities.

Still, not a single journalist, editorial writer, or politician knew the collector's name, where he was from, or anything about the qualifications he brought to his job. Believers in the dossier accepted his bona fides as an article of faith.

The dossier's critics were eager to undercover his identity and, for some time, a small band of internet sleuths had been sifting through clues in government reports and other sources to try to figure out the collector's name. Four of those amateur detectives had stumbled into each other on Twitter and started combining their research and analytical firepower. One member of the group, Hans Mahncke, tweeted under his own name. The rest used

Twitter handles. Walker Hanson, a bank officer, was @Walkafyre. Jimmy Nelson, a postgraduate student in biomedical research, was @Fool_Nelson. A former mining industry executive and climate change skeptic, Stephen McIntyre, was @ClimateAudit.

McIntyre was the only one of the four who had been previously involved with a public controversy and the news media; a 2009 article in *The Wall Street Journal* had described him as "global warming's most dangerous apostate." Ideologically, several of the men were supporters of Donald Trump and rabid believers in a "deep state" conspiracy against him. They were also all obsessives with lots of time on their hands. Their hunt to identify Steele's collector had been underway for a time but it gathered steam when Michael Horowitz, the inspector general of the Justice Department, issued his report in late 2019. In it, there were clues that Steele's collector wasn't "Russian-based," as FBI officials suggested in court filings seeking to conduct surveillance on Carter Page, Trump's ex–foreign policy advisor.

The sleuths began to sense that Steele's collector was living in the United States and might even be a former intelligence agency operative. In early 2020, they wrote a paper naming their suspect and posted it on the internet. They said he was Yuri Shvets, the same former KGB agent with whom Steele had wanted to talk about Alfa Bank.

The amateur detectives, had they known about Steele's 2018 meeting with Van Niekerk, would have dropped Shvets from consideration. But they believed that Shvets and Steele likely crossed paths in connection with the notorious 2006 murder in London of an ex-KGB spy, Alexander Litvinenko. Shvets and Litvinenko had worked together as private operatives prior to Litvinenko's death, and Shvets testified at the official inquest into his murder.

In numerous media accounts, Steele was depicted as spearheading MI6's investigation into the ex-spy's murder, though his precise role wasn't clear. One author of a book about the case, Alan Cowell, said he had never heard Steele's name. Litvinenko's closest friend, Dr. Alexander Goldfarb, said the same thing and added that Litvinenko's widow also hadn't heard of Steele.

The internet sleuths did uncover something about Shvets during their research that should have eliminated him from consideration as Steele's collector. In early 2017, shortly after BuzzFeed posted the dossier, Shvets had given an interview to an obscure Ukrainian outlet and trashed Steele's reports. He pointed out that Steele's claims that his collector had access to multiple Kremlin insiders seemed absurd on its face. But the internet sleuths, given their predilections for conspiracy theories, weren't ready to take Shvets at his word. Instead, they wrote that his disparaging comments about the dossier might have been a "false flag" to steer attention away from him. "Shvets' overall analysis of the Steele dossier was uncannily accurate—far more accurate and far more prescient than any contemporary US observer," their analysis stated. "We cannot help but wonder if, like an arsonist on the scene of a fire he knows more about the dossier than he was letting on."

SEVERAL MONTHS LATER, AS the 2020 U.S. presidential election approached, the guessing game would start again. And this time, it provided an answer.

Christopher Steele's collector, when FBI agents interviewed him in early 2017, was told that the bureau would keep his name secret. But Donald Trump's compliant Attorney General William Barr decided to release those interview notes.

The bureau's discussions with Steele's collector had taken place over a three-day period in late January 2017, soon after BuzzFeed posted the dossier, and were memorialized in a fifty-seven-page memo. As with many government documents, the memo contained lots of redactions that blanked out words or portions of sentences that officials considered too sensitive for public disclosure. Those redactions included, among other things, the name of Steele's collector, the place where he had grown up, the colleges that he had attended, and other information that could have given away his identity.

When those interview notes hit the internet, Carter Page and Sergei Millian quickly tweeted out that they were now sure that Steele's collector was Ed Baumgartner, a Russian-speaking operative who was a contractor for Fusion GPS. They were wrong but Baumgartner, who posted politically snarky comments on Twitter, quickly deleted his account after receiving threats. Meanwhile, the same internet sleuths who mistakenly had pointed a finger at Yuri Shvets were fly-specking the FBI memo. Walker Hanson, who tweeted as @WalkaFyre, had developed a methodology that he used to try to figure out the word or phrase masked by a redaction. As a starting point, he measured the length of a redaction and compared it to a similar length of visible text. That provided a rough count of the number of hidden characters but there was a variable that Hanson had to take into account. It was the typeface, or font, used to create the document. In some typefaces, known as monospace fonts, each character takes up the same amount of space. But there are popular typefaces such as Times New Roman where the size of characters differed markedly and in which a short word covered by a redaction might contain anywhere from four to six characters.

The FBI interview of Steele's source was composed in a typeface called Courier, a monospace font, which made it easier to decipher. And since it contained both the full name and the last name of Steele's collector, Hanson and his fellow detectives had a jumping-off point. The collector's full name appeared to consist of fourteen characters, while his last name contained nine characters. That meant that his first name, accounting for a space between his first and last names, consisted of four characters yielding a formula for his full name of 4+1+9. Something else jumped out from the FBI's interview of the collector. He seemed relatively young given the length of his work history and that ruled out a seasoned ex-operative like Yuri Shvets. The FBI interview notes also indicated the collector had grown up in a Russian city whose name was composed of four characters and had gone to graduate school in the United States at a university with ten characters in its name.

Based on their research, the sleuths were drawn to an energy industry expert, Ilya Zaslavskiy, who had investigated Russian corruption. But just as the detectives were running down clues, someone beat them to the punch. An anonymous blogger posted a site called ifoundthepss.blogpost.com (in FBI notes, the collector was referred to as Steele's Primary Sub-Source, or pss). And on it, he identified Steele's collector as someone no one had ever thought about or heard about. His name was Igor Danchenko and his first and last names, counting the space between them, fit the 4+1+9 formula.

The creator of the ifoundthepss blog wrote that he had initially suspected that Steele's informant might be one of several young former analysts at Orbis Business Intelligence who had left the firm to start their own business. That guess hadn't panned out but he then started looking at people who followed one of those analysts on Twitter. That's where he stumbled onto Danchenko.

The blogger wrote that he then searched Danchenko's Twitter account and LinkedIn profile and the clues fell into place.

Danchenko had posted pictures of trips to Russia and London at times when Steele's collector would have been there. He had grown up in Perm (four letters), a city in the mining region of Siberia, and had attended graduate school at the University of Louisville (ten letters). He also lived in Washington, D.C., for a time and once worked at the Brookings Institution, a think tank once headed by one of Steele's closest American friends, Strobe Talbott. His career highpoint before the dossier—and presumably one for which Christopher Steele thought he deserved a medal—was Danchenko's discovery that Vladimir Putin had plagiarized parts of his thesis. For conspiracy buffs, the ifoundthepss blogger offered a bonus. At Brookings, Danchenko had worked with Fiona Hill, the Russia expert who had testified about the dossier during the congressional impeachment hearings in 2019.

"Traditional journalism is dead," the blog's author wrote. "If you find the right people online they will tell you a story hours, days, weeks, or even years before mainstream media will deliver it to you. By the time it does get to you it has been filtered and spun to suit whatever agenda and narrative they were pursuing at the time."

That might be true and the person who wrote the ifoundthepss blog might have been an investigative genius. Then again, people inside the U.S. government who wanted to out Danchenko might have given him help.

IGOR DANCHENKO BELONGED TO the same subset of operatives as Rob Moore and Mark Hollingsworth, someone with special skills—in his case, languages—who had found his way into the private spying business as a last resort. He said during his FBI interviews that he had gotten a law degree in Russia and worked first as a mining industry lawyer before deciding to attend graduate school in the United States. He received two master's degrees in international affairs, one from the University of Louisville and the other from Georgetown University, which led to a junior position at the Brookings Institution. But his ability to advance further at the think tank was stymied by his lack of a Ph.D. and, around 2010, he began drifting into the private intelligence business.

Danchenko principally worked as an analyst preparing "political risk" reports for companies doing business in Russia or Eastern Europe. The reports were based on "open source" research—newspaper articles, published studies, etc.—rather than material gathered from informants. Then, in 2010, Danchenko had to scramble to find work after the company where he was employed went bankrupt and a mutual acquaintance introduced him to Christopher Steele.

Danchenko told the FBI that Steele, who had then just started Orbis Business Intelligence, initially give him a small trial job to write a report about business risks in Eastern Europe. Steele "liked" his work and, afterward, Danchenko signed a contract with Orbis. His early assignments involved standard analytical research. But in 2012, Steele paid him to take his first trip to Russia as an operative to gather information about a businessman's possible ties to Russian organized crime.

The former MI6 agent told him that for security purposes he shouldn't take notes of his interviews with his sources and should never contact Orbis from Russia, absent an emergency. Danchenko said that after he began regularly traveling to Russia for Orbis he would meet with Steele in London on his way back to the United States and give him verbal briefings about what he had learned. He had heard Steele once worked as a spy, he said, but decided not to ask him about it because he wanted to "stay out of government's business." At that point in his FBI interview, Danchenko's lawyer quipped, "Well, you haven't done a very good job of that."

IT WAS IN MARCH 2016, Danchenko said, that Steele asked him
to start digging up dirt on Paul Manafort, including finding out
about Manafort's activities in Ukraine in the mid-2000s and any
corrupt business deals in which he had been involved. Danchenko,
who was in his late thirties at the time, told the FBI he was
thrown by the assignment because he was "clueless" about who
Manafort was and because Steele's request involved an American. Three months later, in June 2016, when Danchenko went
to Russia for Orbis on a business-related assignment, Steele told
him to also look for any political or business dirt about Donald
Trump and other people, including Carter Page, Michael Cohen,
and Michael Flynn.

Danchenko explained to the FBI that the information he
gathered for the dossier came from a small network of contacts he
had tapped while working on other Russian-related assignments
for Orbis. Based on his descriptions they didn't seem wired into
the Kremlin. They were mainly Danchenko's childhood friends,
college buddies, or drinking partners. One of them, who shared
political gossip with Danchenko, regularly pitched him about getting together on moneymaking ideas. Another friend who came
into contact with Russian government officials through his work
told Danchenko during drinking sessions about the scuttlebutt he
had heard. He would say, "I just heard this from a deputy minister," or "I just overhead such-and-such about an issue," Danchenko
told the FBI.

AFTER BUZZFEED PUBLISHED THE dossier in 2017, journalists
said that Glenn Simpson described Sergei Millian, the real broker claiming ties to the Trump Organization, as the unwitting
source for some of its most explosive claims in the reports and
identified him as "Source D." Three years later, Luke Harding of

The Guardian, a journalist who appeared to serve as Christopher Steele's Dropbox, was still pushing that narrative. "Steele considered how to approach Millian, a potential useful source," Harding wrote in his 2020 book, *Shadow State*. "A direct attempt wouldn't work, so Steele sent an intermediary. Millian spoke at length and privately to this person, believing him or her to be trustworthy—a kindred soul. The conversation was relayed to Steele. Millian was the main source for the dossier's most eruptive claim: that Trump was filmed with Russian prostitutes in November 2013 while staying at Moscow's Ritz-Carlton Hotel."

However, Steele's collector—Igor Danchenko—told a far different story in 2017 to the FBI and it wasn't a tale of kindred spirits. He said his efforts to use the ruse of a potential real estate deal as a way of luring Millian into a meeting had failed and he told FBI agents that he had never met the broker or spoken to him in person. He thought he talked briefly by phone with Millian on one occasion for ten minutes but he wasn't even sure about that because the person who he thought might be Millian hadn't identified himself or said anything particularly revelatory.

DANCHENKO TOLD THE FBI that the real source for some of the dossier's most explosive claims was a woman who had been a key source for many earlier jobs he had done for Orbis. He described her as a childhood friend whom he had known since middle school years in Russia and added he had "100%" confidence in her material.

After Danchenko's interview was posted online, it didn't take long for internet sleuths to locate a likely suspect. Her name was Olga Galkina and her Facebook pages showed that she had worked over the years as a journalist and as a spokesperson for a variety of government agencies and companies. As it turned out, one of those companies was the internet service provider whose owner would sue Steele and Orbis for defaming him.

According to Danchenko, Galkina was the sole source of the material about Michael Cohen's supposed trip to Prague. She first told him in the summer of 2016 that she had heard that Cohen and "three other unidentified people" had flown into Prague to meet with representatives of the Kremlin, he told the FBI. Initially, she said she didn't know the identities of the Kremlin representatives, adding it was her "best guess" they were from the "Legal Affairs" or the "Legal Department," an agency that Danchenko said he had never heard of and didn't think existed. Then, Galkina provided a different story about the Prague meeting. And it was in that version that she said Cohen had met in Prague with a top official of Rossotrudnichestvo, the Russian cultural group. Danchenko told the FBI that he had pressed her for the names of other American participants who had supposedly taken part in the Prague meeting but said she couldn't find out their identities. Several months after the FBI's meeting with Danchenko, agents tracked down Galkina and interviewed her. Not surprisingly, she backed away from what Danchenko had said she had told him and later claimed she was not a source for the dossier.

AS FOR THE "PEE tape," Danchenko explained that one of his sources told him there was a "well-known" allegation that Trump was into "water sports" and had indulged in that predilection at the Ritz-Carlton. When Danchenko spoke to managers at the hotel, they were bemused, saying that "all kinds of things happen" there and that with celebrities "one never knows what they're doing." A hotel housekeeper added that "anything goes" at the Ritz-Carlton, though officially, she added, "we don't have prostitutes." But there was a basic problem with the "pee tape" story that should have been clear from the start to a seasoned spy such as Christopher Steele. Blackmail works best when only a few people know about it—those in possession of the blackmail and the person who is susceptible to it. The Kremlin might have had a "pee tape" on Trump, but if it did it was unlikely it would have been the talk of Moscow.

THERE'S LITTLE QUESTION THAT Igor Danchenko, in speaking with the FBI, had an incentive to cast himself in the best light. Also, there was no independent way to resolve the discrepancies between his story and the information contained in the dossier. Danchenko insisted that he had always been cautious when describing to Steele the quality of the information he had gathered from informants by using phrases such as "as analyst, I think," or classifying material such as "possible vs. likely." Steele's memos didn't contain a hint of evocation. That might have reflected his confidence in Danchenko. It also might have reflected the bravado of a private spy who wanted people to believe he knew more than he did.

The corporate investigator in London, Andrew Wordsworth, who had earlier told Glenn Simpson that he thought Steele's reports were "shit," refused at first to believe that the former spy had employed Igor Danchenko as his collector. Wordsworth said he had known Danchenko for years, thought he was a very nice guy, and hired him occasionally to work on "political risk" reports.

But he added that he never would have assigned Danchenko to gather sensitive intelligence because it was clear to him that he didn't have sources inside the Kremlin or anywhere near it. "Igor is a really nice guy and the thought of [Christopher] Steele presenting him as a super source is beyond embarrassing," he said.

Steele believed, as others did, that Moscow posed a threat to the United States and Europe. But despite those stakes, he seemingly pursued his Trump/Russia investigation as just another job on his plate. According to Danchenko, Steele never met with any of his informants to double-check on what they were telling him. That might have been understandable in the case of sources who were in Russia. But Danchenko's key contact, Olga Galkina, who supplied the critical allegations about Michael Cohen and other matters, lived in Cyprus, and while the FBI took the trouble to meet with her, Steele didn't.

YEARS LATER, AFTER HIS role in the dossier became public, Igor Danchenko gave conflicting interviews about it. He told Luke Harding of *The Guardian* that he was confident about the material he passed on to Christopher Steele including the "pee tape" story. "I got it right," he insisted. But in an interview published on the same day in *The New York Times*, he said he remained skeptical about claims in Steele's memos.

Whatever the case, Danchenko's role in the dossier would bring certain issues into focus. Chief among them was the pedestrian nature of the work performed by operatives-for-hire. Private spies prosper because they operate behind a façade, one that masks the quality of the "intelligence" they sell to clients from scrutiny. That secrecy is the key to the *Wizard of Oz* nature of the corporate investigations industry. As long as their work never becomes public, operatives can claim to customers that they are selling them "strategic intelligence" when what they are often doing is selling smoke. It becomes plain after the smoke clears that private spies don't just play their targets. Their customers can get played, too.

IN 2020, THE U.S. Senate Select Committee on Intelligence, which had been investigating the Trump/Russia issue for three years, issued a long-awaited report. Unlike much of the political posturing that was a hallmark of the Trump era, the panel's bipartisan finding confirmed that Moscow had sought to influence the 2016 campaign on Trump's behalf. In their report, lawmakers also addressed the dossier and went to great lengths to point out that there was one major player in Kremlin-connected influence operations whose name had been conspicuously absent from all of Christopher Steele's memos. It happened to be the same oligarch on whose behalf Steele had worked for five years as a private operative—Oleg Deripaska.

When the Trump/Russia controversy exploded, the Associated Press reported in 2017 that Paul Manafort had sent a $10 million proposal a decade earlier to Deripaska, offering to do lobbying work to advance Kremlin interests. Deripaska never signed the deal and offered to provide information about Manafort to congressional investigators.

The U.S. Senate Select Committee on Intelligence decided that it could make do without the oligarch's help, and Steele repeatedly insisted that his ties to Deripaska hadn't influenced what was in the dossier. Still, the senate panel pointed out in its report that it was extremely curious that the dossier never mentioned Deripaska given the oligarch's efforts over the years to advance the Kremlin's political objectives. "Steele mentions Paul Manafort by name roughly 20 times in the dossier, always in the context of his work in Ukraine; and, in particular, Manafort's work on behalf of then Ukraine President Victor Yanukovych," the panel wrote. "Deripaska, who had a long-standing business relationship with Manafort, is not mentioned once."

CHRISTOPHER STEELE'S TIES TO Deripaska ended in 2017 but he was not the only hired operative who hadn't broadcast his connections to the oligarch to journalists or lawmakers. The Senate Select Committee on Intelligence noted in its report that Glenn Simpson hadn't discussed his links to the oligarch during his testimony before the panel. As it turned out, Steele hired Fusion GPS in March 2016, three months before Simpson retained him to produce his Trump/Russia memos, to gather material about Paul Manafort, apparently to aid Deripaska's lawsuit against the lobbyist.

BY 2020, THE NEWS media had largely lost interest in the dossier but Fusion GPS was still dealing with fallout from it. The lawsuit filed in the United States by the principals of Alfa Bank against the firm was proceeding. Fusion GPS insisted it had done nothing wrong. The oligarchs' lawyers were also seeking to depose Glenn Simpson, Dan Jones, and other people involved with the "pinging" stories.

Glenn Simpson had a last hurrah. From the beginning of his career as a private operative, he had been eager to have his work chronicled on film. In 2009, he had pitched Brian Ross and Rhonda Schwartz of ABC News on the idea of doing pieces based on his cases. A decade later, Fusion GPS approached a film company run by Alex Gibney, the well-known documentary filmmaker, with a similar proposal.

The company passed. But not long afterward, Alex Gibney and Lowell Bergman, the former *60 Minutes* producer, started working on a documentary about Russian election interference. Simpson told Gibney that he was desperate to tell the story of the dossier as part of the film so it would be on tape in case someone tried to harm him.

In March 2017, Gibney traveled to what he later described in the documentary as an "undisclosed" location in California to interview Simpson. In it, the former journalist appears on screen looking like someone in a witness protection program. "I was becoming concerned that if I didn't tell some people this story and something did happen to me . . . um . . . that it would never get out," Simpson says.

Simpson may have been over his worst fears by then because he and Dan Jones had already been out raising funds for the Democracy Integrity Project. Later in the same documentary, Simpson is interviewed by Lowell Bergman in the offices of Fusion GPS.

"So, you're not a journalist?" Bergman asked him.

"I'm not a journalist," Simpson replied.

"You're a gun for hire," Bergman said.

Simpson smiled tightly and responded, "Call me what you want . . ."

AFTERWORD

In 2010, Eamon Javers published a book about the corporate investigations industry titled *Broker, Trader, Lawyer, Spy*. In his conclusion, Javers suggested that one way to rein in the renegade activities of such firms would be to impose the same type of registration and disclosure requirements that Congress requires of lobbyists who are paid to influence lawmakers. He wrote that a regulatory body, the Securities and Exchange Commission, could set up a kind of "spy registry" in which operatives-for-hire would have to disclose the names of their clients and assignments. "It's time for the spy firms to come in from the cold," Javers wrote.

It was a well-intended proposal. The problem is that during the decade between the appearance of that book and this one, the behavior of corporate intelligence firms has only become more predatory and abusive. Few professions are more deserving of a comeuppance than the smug and morally bankrupt men and women who work as spies-for-hire. Suing them, wherever and whenever possible, might help trim their sails but only if plaintiffs in those actions are willing to go to the mat and, rather than taking money to go away, force the truth to come out.

Law enforcement authorities could help by bringing criminal charges against rogue private spies and their sponsors. Following the uncovering by Citizen Lab of the "hacking-for-hire" operation in India, federal prosecutors arrested an Israeli-based private investigator and charged him with farming out work to the hackers. But given the breadth of the hacking operation, the Israeli operative's role in it was likely minor, and it wasn't clear whether law enforcement authorities were prepared to send a real message to the industry by going after the beneficiaries of the hacking and the lawyers or operatives who helped arrange for it.

In the end, the best defense against these mercenaries may be self-defense. The corporate investigations industry presents itself as a technologically-savvy twenty-first-century business but, at its heart, it remains a con game in which we are all potential marks. The old British private detective, Ian Withers, who had helped me stalk Christopher Steele, understands far better than most people how the game is played because he has played it for decades. And his advice, like that of John Scott-Railton, the cybersecurity expert at Citizen Lab, boils down to a simple instruction: don't be a sucker.

"The golden rule applies in all kinds of scams," Ian said. "Never give any information over the phone to a caller you don't know."

One solution, he added, is to tell an unknown caller to send you an email stating the reason they are contacting you as well as a phone number at which you can reach them. But that number, Ian warns, could be a dummy phone that bounces calls to another phone, so it's good to put any number you are given into Google and see what comes up. In addition, there are a variety of public and inexpensive databases where one can determine if a company described by a stranger as their employer exists and, if it does, how recently it was formed. Had people approached by Black Cube operatives taken those steps they could have saved themselves grief.

An operative may also start trying to groom a target by approaching them about an issue that has nothing to do with their real objective, as the Black Cube agent did with one of Harvey Weinstein's accusers. "Don't fall into the trap," Ian explained. "The trap will be very loose. Someone asks for your opinion but that leads to the relationship building in your mind."

There is one profession that can impact how private spies operate—my own, journalism. Reporters often mask the identity of people who they draw upon as sources by using vague, descriptive terms such as "a source familiar with events" or a "knowledgeable source."

A while back, *The New York Times* and other major outlets changed policies and now require more detailed descriptions about anonymous sources so readers can judge for themselves if they have an axe to grind. Media organizations made those changes to foster transparency, a virtue to which every journalist says they subscribe. But to the private spying industry, which profits from secrecy and deception, transparency is anathema.

When Glenn Simpson and Peter Fritsch promoted the dossier, they made it clear that if a journalist wanted to play ball with them, he or she had to agree to keep secret their names as well as Christopher Steele's. That deal became the basis for pacts that allowed the dossier story to flourish for years. (While I was at *The New York Times*, I spoke with Glenn Simpson on several occasions, though I don't recall writing anything based on our discussions and I didn't know anything about the dossier until BuzzFeed published it.)

What made the agreements that Simpson and Fritsch struck with journalists all the more remarkable was that they weren't remarkable at all. Journalists, hungry for stories and headlines, have allowed private spies to set the rules of engagement between them for decades.

There is nothing wrong with private operatives approaching journalists with information they have gathered. But where things do go wrong—and where the public is badly misled—is when news organizations allow the involvement of hired operatives in a piece of journalism to remain concealed. If anyone has a bias about the subject of a story or an interest in slanting information to favor their client, they do. That's why they get paid. Yet media organizations provide them with the same degree of anonymity and protection they rightfully give to a whistleblower who is risking his or her career or personal safety by speaking out.

It isn't clear why journalists keep playing this game. Perhaps media outlets fear that corporate intelligence firms (or their fellow enablers in the legal or crisis management trade) will cut them off if they start setting the rules. That is a small price to pay, particularly given the fact that many private operatives who will feed a story to a reporter on behalf of one client will then turn around and spy on them for another client. Readers and viewers have a right to know when material that appears in a piece originates with an operative paid to dig it and plant it. The problem isn't unique to mainstream media outlets. Conservative publications routinely serve as dumping grounds for all kinds of trash, including garbage claims of voter fraud after Donald Trump's loss in the 2020 presidential election.

Look, I get it. I was a reporter for decades and I loved scoops, too. But if journalists don't break their pact of secrecy with private spies it's only a matter of time before another debacle like the dossier happens again.

ACKNOWLEDGMENTS

Writing is a solitary pursuit. Turning an idea into a published book is anything but solitary. It happens because a large number of people generously share their talents and time. This was particularly true for books like this one which took shape amid the hardships and horrors of a global pandemic.

My agent, Farley Chase, worked from the start to help shape the proposal for this book and found a wonderful editor in Jonathan Jao at HarperCollins. From there, Jonathan's good humor and enthusiasm for this project took over and propelled it forward. Farley's co-agent in London, Caspian Dennis, connected with Juliet Brooke of Sceptre Books, who was eager to bring out a UK edition of *Spooked*. Juliet brought her own sensibilities to the manuscript, and the final results benefited from them. It was a thrill, as well as a first for me, to have a book published simultaneously on both sides of the Atlantic. That may also mean that I forget to acknowledge everyone at HarperCollins and Sceptre who helped to bring this book into print and, if so, my apologies in advance.

At HarperCollins, props go to Sarah Haugen for all the hand holding she provided. Nancy Singer created a clean, pleasing design for the book's interior, and Richard Ljoenes designed a sparkling (and spooky) cover. Kyran Cassidy applied his legal eye to the manuscript, and Becca Putnam handled marketing. My thanks go to Theresa Dooley, who worked fervently to bring attention to the book.

At Sceptre, shout-outs go to Irene Rolleston, who kept me abreast of the book's progress; Helen Flood, who handled marketing; and Maria Garbutt-Lucero, who worked to generate publicity and media interest in the book. Kirsty Howarth read the book with a particular eye to the nuances of British law. Major thanks also go to Lewis Csizmazia, who designed the cover for the British edition of *Spooked* but who also created a striking cover for my first book, *Pain Killer*, which Sceptre brought out in late 2020. The author photograph for the jacket was taken by my former *New York Times* colleague and friend Peter Eavis, who was ideal for this transatlantic assignment since he is a Brit and a diehard Liverpool fan who lives in New York. I first reported for *The New York Times* about Rob Moore and K2 Intelligence, and the newspaper was kind enough to allow me to draw upon my notes.

Books of contemporary nonfiction can happen only if people are willing to take time to speak to an author. I wish everyone I approached had agreed to talk to me and fortunately many of them did. Some asked that I not name them in the book. They know who they are and I am grateful for their participation.

What follows is an alphabetical, unweighted listing of who shared their experiences, insights, thoughts, or provided support and help. They include, among others: Jill Abramson, Rinat Akhmetshin, Rupert Allason, Laurie Kazan-Allen, Marie Arana, Daniel Balint-Kurti, Marty Baron, Dean Baquet, Paul Barrett, Lili Bayer, Susan Beachy, Rich Behar, Kitty Bennett, Ken Bensinger, Ronen Bergman, Lowell Bergman, David Boies, Simon Bowers, Val Broeksmit, Bill Browder, Clare Rewcastle Brown, Oliver Burroughs, Ambrose Carey, John Carreyrou, Barry Castleman,

Julie Cohen, Laurie Cohen, David Corn, Alan Cowell, Shawn Crispin, Alan Cullison, Ken Dilanian, David Enrich, Elena Egawhary, Howard Chua-Eoan, Matt Flegenheimer, Charles Francis, Pierre Gastineau, Jeff Gerth, Simon Goodley, Alex Gibney, Alexander Goldfarb, Adam Goldman, Juleanna Glover, Patrick Grey, Bill Grueskin, Brian Gruley, Walter Hanson, Andrew Higgins, Mark Hosenball, Jeff Katz, Arpad Krizsan, Andrew Kramer, Nicole Hong, Scott Horton, Adam Hulcoop, Michael Isikoff, Eamon Javers, Andrew Jennings, Mark Landler, Amy Lashinsky, Eric Lichtblau, Jane Mayer, Mark Mazzetti, David McCraw, Stephen McIntyre, Jim Mintz, John Mintz, Alexander Mirtchev, Rob Moore, John Moscow, Conrad Mulcahy, Steven Lee Myers, Jimmy Nelson, Igor Ostrovskiy, Nick Peck, William Pigman, David Plott, Matt Purdy, John Scott-Railton, Matthew Rosenberg, Brian Ross, Chuck Ross, Rebecca Ruiz, Raphael Satter, Diana Schemo, Rhonda Schwartz, Scott Shane, Ken Silverstein, Cam Simpson, Ben Smith, Abigail Fielding-Smith, Jay Solomon, Paul Steiger, Simon Taylor, Megan Twohey, Josh Tyrangiel, Phillip Van Niekerk, Frank VanderSloot, Tapio Vaskio, Natalia Veselnitskaya, Erik Wemple, Franz Wild, Ian Withers, John Withers, Andrew Woodsworth, and Jim Yardley.

Like most people, I didn't get to see enough friends and neighbors over the past year due to the pandemic. But even short exchanges over Zoom or email helped cheer me up. So a big thanks to Marty Baron, Alice Black, Susan Bernfield, Eric Chinski, Louise Crandell, Marie Costello, Dick Einhorn, Neill Fernley, Sam Grobart, Eve Hahn, Steve LeVine, Mark Liebling, Cheryl Liebling, Andy Martin, Marshall Messer, Jim Mintz, Jad Mouawad, Michael Moss, Claude Millman, Phil Parker, Kristin Powers, Hilary Redmon, Ellen Rooney, Rebecca Ruiz, Becky Saletan, Gene Shafer, Amy Singer, Deborah Stewart, and David Udell. Our next-door neighbors—Kenny Powers and Lorrie Yowell—were lifesavers over the past year. Then, there were friends old and new in Tyringham, Massachusetts. And despite Covid, it was another wonderful summer for Monterey softball.

This has been a difficult year for so many people. Many have lost loved ones. Others have suffered illness. Still others have braved dangers to serve the rest of us. I hope we never forget these sacrifices.

Nothing would be possible for me without the love and tolerance of my daughter, Lily; my wife, Ellen; and our dog, Charlie, the Havanese. I wondered for a time how I would mention Ellen in this book's acknowledgments. Throughout much of its writing, my wife, who is an extraordinarily talented editor, wanted to have nothing to do with it. Her position was simple. She said she had invested way too much of her time editing my other books. She was right. But then, just as this book was nearly done, she read it and made a number of fantastic suggestions. She said she thought it was "among my top three books." (I've written two others.) That's how Ellen rolls, and I'm lucky to have a spouse who keeps me laughing, even during these darkest of times.

NOTE ON SOURCES

This book was based on more than 130 interviews conducted over a two-year period with people in the corporate investigations industry, the news media, and other professions. A large number of them spoke on the record but others agreed to speak only on background, for fear of alienating their professional colleagues or customers. This book draws on various court filings both in the United States and England. With respect to Mark Hollingsworth's emails, some of those I cited are found in court filings made by Eurasian Natural Resources Corporation while others are contained in the dump of his emails. For the readers' benefit, I took the liberty of capitalizing words in Peter Fritsch's emails.

NOTES

PROLOGUE: STALKING MR. STEELE

1 MI6: The Secret Intelligence Service, the United Kingdom's overseas intelligence agency, is commonly known as MI6.

3 a new editor of *Vanity Fair*: The editor was Radhika Jones.

3 published a profile of Steele: Howard Blum's profile of Steele appeared in the April 2017 issue of *Vanity Fair*.

4 a woman who claimed to be a journalist: This incident was reported by Agence France-Press in a May 18, 2019, article, "Consultant poses as journalist in Monsanto trial."

4 locked in a dispute with three businessmen: This claim by Neil Gerrard is contained in a 2019 lawsuit he filed against Diligence International and ENRC.

4 The chase ended with: In published reports, the investigative firm involved in the Geneva incident denied any wrongdoing and Credit Suisse said the news media had sensationalized the incident.

5 a book about a former FBI agent: My book about Robert Levinson, *Missing Man*, was published in 2016 by Farrar, Straus & Giroux.

6 When I approached Glenn Simpson: He also did not respond to a series of questions sent to him during the reporting for this book.

6 Neither would his partner: Peter Fritsch also did not respond to written questions sent to him during the reporting for this book.

6 A British journalist described Ian: The journalist was Nick Davies in his book *Hack Attack*.

6 "Spies Island": The *60 Minutes* episode in which Ian Withers appeared was broadcast on March 26, 1989.

7 Jane Mayer, in a profile of Christopher Steele: Mayer's piece appeared in the March 12, 2018, issue of *The New Yorker*, "Christopher Steele, the Man Behind the Trump Dossier."

11 "We do not intend to respond": email from Arthur Snell of Orbis to me, October 12, 2020.

CHAPTER 1: JOURNALISM FOR RENT

16 One ex-colleague dubbed him "Shaggy": Dana Milbank described Simpson in a January 12, 2018, column he wrote for *The Washington Post*, "I Know Glenn Simpson. He Is Not a Hillary Clinton Hit Man."

16 Simpson's desk was a thicket: The description is drawn from Matt Flegenheimer's article in *The New York Times* on January 8, 2018, "Fusion GPS Founder Hauled from the Shadows for the Russia Election Investigation."

17 to buy the *Journal*'s parent company: The newspaper was owned by Dow Jones & Company.

18 *Journal* reporters felt whipsawed: David Carr, the late media editor for *The New York Times* chronicled concerns expressed by Journal reporters in a piece published on December 14, 2009.

21 that mentioned Browder's expulsion: That article, "How Russian Tycoons Got Wealthy," was published on March 28, 2006.

22 One was Oleg Deripaska: An article by Glenn Simpson and Mary Jacoby that mentioned Deripaska, "How Lobbyists Help Ex-Soviets Woo Washington," was published in *The Wall Street Journal* on April 17, 2007.

22 The other was a well-known Washington lobbyist: An article by Glenn Simpson and Mary Jacoby that mentioned Paul Manafort, "McCain Consultant Is Tied to Work for Ukraine Party," was published in *The Wall Street Journal* on May 14, 2008.

23 When Simpson started writing about Kazakhstan: In 2008, Simpson and his reporting partner would write a series of articles for *The Wall Street Journal* about the country's president, Nursultan Nazarbayev, his ex son-in-law, Rakhat Aliyev, and a Kazakh-connected consultant in Washington, D.C., Alexander Mirtchev. The articles included "Kazakh Leader's Daughter Monitored U.S. Bribe Case," May 12, 2008 (on which Mary Jacoby was

a coauthor); "Kazakhstan Corruption: Exile Alleges New Details," July 22, 2008; "Perle Linked to Kurdish Oil Plan," July 29, 2008; "Oilman Protests Prosecution Delay," September 25, 2008; "Russia's Deripaska Faces Western Investigations," October 10, 2008.

23 Rakhat Aliyev: In 2015, he was found dead in a prison cell in Austria. His death by hanging was ruled a suicide.

23 the U.S. Justice Department charged an American consultant: That consultant was James H. Giffen. He asserted at this trial that he was acting with the approval of CIA. Most charges against him were dropped and he pleaded guilty to a minor tax charge. Nursultan Nazarbayev, the Kazakh president, denied he had received bribes.

26 dubbed "Operation Hellenic": I obtained several interim reports produced during Operation Hellenic that were dated September 2007, October 2007, November 2007, and January 2008. Several corporate investigators were unable to identify the firm that produced them.

27 In a subsequent report: "Operation Hellenic" report dated November 2, 2007.

29 a 2009 State Department cable: The cable, which is dated January 13, 2009, was among the trove of documents disclosed by WikiLeaks. The U.S. ambassador in Kazakhstan was then Richard Hoagland.

29 with the help of Baker and Hostetler LLC: This State Department document was released by Wikileaks. Over the years, Baker and Hostetler has used differing styles for its corporate name. It currently uses BakerHostetler and I have used it throughout.

CHAPTER 2: "LAPDANCE ISLAND"

33 When a freelance private spy named Rob Moore: I first wrote about Rob Moore's work for K2 Intelligence in *The New York Times* on April 27, 2018, "A Spy's Tale: The TV Prankster Who Says He Became a Double Agent."

34 "Lapdance Island": a short clip from the show survives, https://www.you tube.com/watch?v=ez5WYdGHoCw.

36 "Commercial Spies Tap State Records": This *Guardian* article was published on May 11, 1971.

37 Mary Cuddehe, described how the Kroll official: Her article about Kroll's efforts to recruit her appeared in *The Atlantic* magazine on August 2, 2010, "A Spy in the Jungle."

39 "Spider-Man, the Hulk—I really owe my business": This anecdote comes from William Finnegan's profile of Jules Kroll that appeared in *The New Yorker* on October 19, 2009, "The Secret Keeper."

40 T. Boone Pickens, complained that Kroll Associates: The corporate raider was quoted by Douglas Frantz in a *Los Angeles Times* article of October 9, 1988, "Kroll—Wall Street's Super Sleuth."

40 "Wall Street's private eye": This description of Jules Kroll was contained in a March 4, 1985, article by Fred Blakeley in *The New York Times*, "Wall Street's Private Eye."

40 "Many of us are corporate misfits": That lawyer, Charles E. Bohlen Jr., was quoted in that same *Los Angeles Times* article.

41 a future U.S. president, Donald J. Trump: The anecdote about Jules Kroll's dealings with Trump over an Atlantic City casino is drawn from an article by Christopher Byron in *New York* magazine, "High Spy," published on May 13, 1991.

42 But an article in *New York* magazine: This discrepancy was noted in Christopher Byron's 1991 article "High Spy."

42 Hakluyt used a similar tactic: An account of the investigative firm's infiltration of Greenpeace appeared in *The Sunday Times* (of London) on June 17, 2001. In that account, Hakluyt declined to comment and Shell and BP said they were not aware of the strategies used by investigators.

43 He was a German-born spook: The name of the operative who infiltrated Greenpeace was Manfred Schlickenrieder.

43 Diligence, got caught: A detailed account of Diligence's escapade, which I summarized, was written by Eamon Javers for *BusinessWeek* magazine and appeared on February 26, 2007, "Spies, Lies and KPMG."

43 according to an account in *Bloomberg Businessweek* magazine: That piece by Eamon Javers, "Spies, Lies and KPMG," was published on February 6, 2007.

44 his firm got sued: Diligence insisted that all its activities were legal. It was sued by both KPMG and the investment fund, IPOC International Growth Fund Limited. All the legal cases were settled or resolved without any acknowledgement of wrongdoing.

46 A law firm fired Kroll Associates: This incident and other problems Kroll encountered in the 1990s were chronicled by William M. Carley of *The*

Wall Street Journal in a November 17, 1994, article, "Kroll Calling: Big Detective Agency Finds Sleuthing Tougher in the '90's."

46 Charges in Brazil against several Kroll employees including Charlie Carr were dropped in 2012 for lack of evidence, according to published accounts.

46 One former FBI agent told *Vanity Fair* magazine: The account of Kroll's work for R. Allen Stanford was detailed by Bryan Burrough in a June 22, 2009, article, "Pirate of the Caribbean."

47 "pure cockroach": Stanford's instructions to a Kroll operative were first disclosed by reporters for McClatchy in a November 29, 2012, article, "Ponzi Artist Investigated Ex-State Department Official, Records Show." The Kroll executive who undertook that assignment was Tom Cash, a former DEA agent.

47 A construction industry trade group: The group that sued Kroll Inc. for its alleged failure to disclose its ties to Allen Stanford was the National Electrical Contractors Association.

48 "clearly a blemish": Jules Kroll made those comments to James Freeman of *The Wall Street Journal* in a September 4, 2010, article, "Where There's Corruption, There's Opportunity."

48 merged Kroll with an armored car maker: The 1997 merger involved O'Gara, Hess & Eisenhardt.

48 sold Kroll to a big insurance company: The insurer was Marsh & McLennan.

49 raided by the police: Charges against Carr were dropped due to lack of evidence

53 "Victims of Chrysotile Asbestos": The short film can be viewed at https://vimeo/171175438.

CHAPTER 3: "OPPO"

56 Alex Yearsley: He did not respond to repeated requests for comment during the course of reporting for this book.

56 an obscure company in Ukraine: RosUkrEnergo was the name of the intermediary company involved in the Ukraine natural gas pipeline.

57 a front-page article for the *Journal*: Glenn Simpson's article about Semion Mogilevich appeared on December 26, 2006, "U.S. Probes Possible Crime Links to Russian Natural Gas Deal."

59 "Thank you, America, our people will soon be safe": The description of the sheik's publicity campaign was drawn from an article in *The Guardian* on June 7, 2010, "A Very Peculiar Coup: UK Solicitor in Plot to Take Over Emirate."

59 Simpson was paid $40,000 by a lobbying firm: Simpson's involvement in lobbying on behalf of Sheik Khalid bin Saqr al-Qasimi was disclosed by an article in *The Hill* by Kevin Bogardus on November 4, 2009, "Deposed Sheik Hires Former Reporter in Lobbying Effort."

60 "out of an abundance of caution": Simpson's reason for registering as a lobbyist appeared in the same *Hill* article.

60 his best-selling memoir, *Red Notice*: Bill Browder's book, *Red Notice: A True Story of High Finance, Murder, and One Man's Fight for Justice* was published in 2015 by Simon & Schuster.

62 charged that he had used a racial epithet: Email from Shawn Crispin to Simon Mennell dated December 16, 2004.

64 Isikoff learned that the Democratic National Committee: Michael Isikoff's article about the use of private operatives by the Bill Clinton campaign appeared in *The Washington Post* on July 26, 1992, "Clinton Team Works to Deflect Allegations on Nominee's Private Life."

65 *Dirty Little Secrets*: The book by Larry Sabato and Glenn Simpson, *Dirty Little Secrets: The Persistence of Corruption in American Politics*, was published in 1996 by Times Books/Random House.

65 Terry Lenzner: He died in April 2020 after a long illness.

66 whose company sold dietary supplements: The name of the company was Melaleuca Incorporated.

66 an article in *Mother Jones* magazine: The piece about Frank VanderSloot by Stephanie Mencimer appeared in *Mother Jones* on February 6, 2012, "Pyramid-Like Company Ponies Up $1 Million for Mitt Romney."

67 Strassel wrote a column about the episode: Kimberly Strassel's article about Frank VanderSloot and Fusion GPS appeared in *The Wall Street Journal* on May 10, 2012, "Trolling for Dirt on the President's List."

CHAPTER 4: THE LONDON INFORMATION EXCHANGE

73 Eurasian Natural Resources Corporation: Throughout the reporting of this book, ENRC and its outside media representative did not respond to inquiries or written questions.

73 the "Kazakh Trio" or the "Trio": The three oligarchs who composed the Trio were Patokh Chodiev, Alexander Machkevitch, and Alijan Ibragimov.

73 one titled *Londongrad*: That book, which Hollingsworth coauthored with Stewart Lansley, was published in 2010 by Fourth Estate.

74 his corporate bio stated: Robert Trevelyan was involved with numerous companies. This bio was on the website of one called Luxian.

75 Trevelyan made copies of ENRC hard drives: ENRC made this allegation in numerous lawsuits.

75 One of his supposed partners in that enterprise: ENRC made this claim in a 2019 lawsuit filed in London against Mark Hollingsworth.

75 a corporate intelligence firm hired by the company: The firm was Nardello & Company.

75 then publicly traded: ENRC shares were delisted on the London Stock Exchange in 2013.

76 "In my view, this is pretty devestating": Email from Mark Hollingsworth to Glenn Simpson, July 28, 2011.

76 "I have to brief client orally on Monday": Email from Glenn Simpson to Mark Hollingsworth, November 30, 2011.

76 "Our mutual friend Magic": Email from Mark Hollingsworth to Glenn Simpson, January 29, 2012.

77 he asked a sommelier to serve him: This anecdote is taken from a piece by Katrina Mason in *The Financial Times* on September 1, 2017, "Russian Lobbyist Rinat Akhmetshin on That Notorious Meeting at Trump Tower."

78 a Global Witness investigator: That investigator was Daniel Balint-Kurti.

79 Baruch Halpert: He did not respond to numerous messages seeking comment.

79 IMR had hired a private spying firm to investigate Akhmetshin's suspected role: My description of GlobalSource's investigation of Rinat Akhmetshin is drawn from court filings made in connection with a so-called 1782 proceeding brought by IMR in 2014 against him. Section 1782

is part of a federal statute that allows a litigant in a foreign proceeding to
seek discovery from a person in the United States.

80 "There is a lot of shit": Affidavit of Akis Phanartzis of GlobalSource
dated March 20, 2014.

81 "We requested that they provide us copies": Affidavit of Raphael Rahav of
GlobalSource dated March 21, 2014.

81 28,000 files belonging to IMR: Affidavit of Tadeusz Jarmolkiewicz, general counsel of IMR, dated March 26, 2014.

82 "You know, people often ask me for information": Deposition of Rinat
Akhmetshin filed May 18, 2015.

CHAPTER 5: BAD BLOOD

86 "B&W's tactics aren't unheard of in high-stakes litigation": Suein Hwang
and Milo Geylin of *The Wall Street Journal* reported about IGI's dossier on
February 1, 1996, "Getting Personal: Brown & Williamson Has 500-Page
Dossier Attacking Chief Critic."

86 "Now he is slimy": That quote appeared in Judy Bachrach's profile of Terry
Lenzner, "The President's Private Eye," which was published in the September 1998 issue of *Vanity Fair*.

86 as apparent payback: A spokesman for the Koch brothers disputed Mayer's
account and told the *The New York Times* he considered it "grossly inaccurate".

86 *Dark Money*: Jane Mayer's book *Dark Money: The Hidden History of the
Billionaires Behind the Rise of the Radical Right* was published in 2016 by
Doubleday.

87 Felch charged in the article: Jason Felch's article about Occidental
College's reporting policies appeared in the *Los Angeles Times* on De-
cember 6, 2013, "College Received More Sex Assault Allegations than
It Reported."

87 and had collaborated with him: That former *Los Angeles Times* reporter
was Ralph Frammalino.

88 Felch later said: He made that comment to Ravi Somaiya of *The New
York Times* for an article, "Los Angeles Times Fires Writer of Articles on
College," that appeared on March 16, 2014. Jason Felch did not respond to
emails seeking comment for this book.

88 the *Los Angeles Times* was forced to run a major correction: The correction and editor's note about Jason Felch's piece appeared on March 14, 2014.

89 U.S. authorities were investigating the company: Federal prosecutors would eventually indict a director of Derwick Associates on money-laundering charges and the firm's chief executive remained, at the time of this writing, under criminal investigation.

89 Jóse de Córdoba: He did not respond to emails seeking comment.

89 U.S. authorities were investigating the company: To date, Derwick Associates has not been charged with wrongdoing. At the time of this writing, one of its co-founders, who has denied any wrongdoing, remains under scrutiny, according to a 2019 article in the *Miami Herald*.

90 "First, big congrats on the big P": Email from Peter Fritsch to John Carreyrou, May 3, 2015.

91 "I do know Riedel": Email from John Carreyrou to Peter Fritsch, May 3, 2015.

92 a 2018 best-selling book, *Bad Blood*: John Carreyrou's book, *Bad Blood: Secrets and Lies in a Silicon Valley Startup*, was published in 2018 by Alfred A. Knopf.

93 "Hey, so something came up": Email from Peter Fritsch to John Carreyrou, May 8, 2015.

93 "I'm working on a very serious story": Email from John Carreyrou to Peter Fritsch, May 8, 2015.

94 "Is she a cult leader?": Email from Peter Fritsch to John Carreyrou, May 11, 2015.

94 "They just named a new GC": Email from Peter Fritsch to John Carreyrou, May 8, 2015.

95 "You are talking to the wrong dude": Email from Peter Fritsch to John Carreyrou, June 4, 2015.

95 "what I can certainly do without is patronizing commentary": Email from John Carreyrou to Peter Fritsch, June 4, 2015.

98 "I would like to not mention Carreyrou by name": Email from Peter Fritsch to Russell Carollo, August 2, 2015.

98 In time, Elizabeth Holmes faced: Both Holmes and her partner, Ramesh Balwani, were later charged with fraud. They pled innocent and their trial, at the time of this writing, was scheduled for March 2021.

98 *The Washington Post*, who was doing a piece about Fusion GPS: That piece by Jack Gillum and Shawn Boburg ran on December 11, 2017, "'Journalism for Rent': Inside the Secretive Firm behind the Trump Dossier."

CHAPTER 6: UKRAINE TOMORROW

102 a journalist with an Israeli television show: The name of the show was *Uvda*.

102 the show was doing an episode about Black Cube: In 2019, Black Cube sued the program, *Uvda*, and its chief correspondent in London claiming the episode had defamed the firm. While maintaining that stance, Black Cube dropped the lawsuit in 2020 saying it couldn't prove financial damages, a requisite of British libel law.

103 when the Harvey Weinstein case erupted: In writing about the Weinstein case, I drew on two books published in 2019 that chronicled the episode and the role of private operatives in it. The book by Ronan Farrow of *The New Yorker*, *Catch and Kill: Lies, Spies, and a Conspiracy to Protect Predators*, was published by Little, Brown. The other book by two *New York Times* reporters, Jodi Kantor and Megan Twohey, *She Said: Breaking the Sexual Harassment Story That Helped Ignite a Movement*, was published by Penguin Press.

106 "You have come across Frank Timis' name": Internal document distributed by Black Cube.

106 "It's like creating a play": The Black Cube advisor was quoted in an article by Bradley Hope and Jacquie McNish that ran in *The Wall Street Journal* on June 18, 2019, "Black Cube: The Bumbling Spies of the Private Mossad."

106 two real estate developers: The developers were brothers, Vincent and Robert Tchenguiz.

107 Firtash was indicted: https://www.justice.gov/opa/pr/six-defendants-indicted-alleged-conspiracy-bribe-government-officials-india-mine-titanium.

109 peppered them with questions about George Soros: The saga of Black Cube's operation in Hungary was chronicled by Lili Bayer of *Politico* on July 16, 2018, "Israeli Intelligence Firm Targeted NGOs during Hungary's Election Campaign."

110 "I was never a Bond girl": Stella Penn Pechanac's comments appeared in *The Daily Mail* on November 30, 2019.

110 "I don't feel guilty about anything": Seth Freedman's comments were broadcast by the BBC on January 13, 2020.

111 when Ronan Farrow of *The New Yorker* disclosed it: Farrow's article appeared on November 6, 2017, "Harvey Weinstein's Army of Spies."

CHAPTER 7: TABLE NUMBER 6

113 the Russian-owned real estate company: Prevezon Holdings was based in Cyprus.

116 later posted on YouTube: The scene is still there: https://www.youtube.com/watch?v=ryVavTF6hR0.

117 "Fighting Putin Doesn't Make You a Saint": That article by Jason Motlagh appeared on December 31, 2015.

118 some $530,000 in fees and expenses: This figure was cited by Chuck Ross in an article in *The Daily Caller* that appeared on November 21, 2017, "Fusion GPS Bank Records Show Russia-Related Payments."

120 told Bloomberg News: That article by Stephanie Baker and Irina Reznik appeared on December 21, 2017, "Mueller Is Looking into a U.S. Foundation Backed by Russian Money."

121 "I can prove in court that Browder": Andrei Nekrasov made those remarks in an article by Henry Johnson published in *Foreign Policy* on June 10, 2018.

123 when the panel ruled, her prediction proved true: John Moscow was barred from representing Prevezon in October 2016.

CHAPTER 8: GLENNTOURAGE

127 at a conference of documentary filmmakers and reporters: The conference was called the Double Exposure Film Festival and was sponsored by 100Reporters, a journalism group.

128 He challenged the anti-abortion group: Jackie Calmes wrote about anti-abortion videos for *The New York Times* in "Planned Parenthood Videos were Altered, Analysis Finds" on August 25, 2017.

130 a court in Ghana: Paul Singer and his vulture investing tactics were described by Sheelah Kolhatkar in *The New Yorker*, August 20, 2018, "Paul Singer, Doomsday Investor."

131 seeking data about U.S. military sales to Argentina: These requests are detailed on a spreadsheet maintained by Carollo.

131 negative articles about another hedge fund: That fund was Gramercy Funds Management; a spokesperson for it declined comment.

132 "*The New York Times*, I know they work with Fusion": Silverstein was quoted by Rowan Scarborough in *The Washington Times*, December 10, 2017, "Fusion GPS Tried to Tie Trump to Clinton's Pedophile Pal Epstein as Part of Smear Campaign."

133 wrote an article about Robert LiButti: Michael Isikoff's article was published by *Yahoo! News* on March 7, 2016, "Trump Challenged Over Ties to Mob-Linked Gambler with Ugly Past."

134 "he will be the first big-time developer since Josef Stalin": Jeff Grocott wrote about Trump's ambitions in a 1996 article in the *Moscow Times*, "Trump Places Bet on New Moscow Skyline."

138 "1. He confirmed that there is what he called": Email from Mark Hollingsworth to Christopher Steele, August 2013.

138 a London lawyer representing Oleg Deripaska: That lawyer was Paul Hauser of Bryan Cave.

138 in the event the oligarch won a favorable court judgment: Oleg Deripaska stopped pursuing his lawsuit against Paul Manafort.

139 as informants: A *New York Times* article published on September 1 2018, "Agents Tried to Flip Oligarchs. The Fallout Spread to Trump" described Bruce Ohr's efforts to recruit Oleg Deripaska.

140 The get-together between Deripaska and Bruce Ohr didn't go well: The meeting was described by Kenneth Vogel and Matthew Rosenberg of *The New York Times* on September 1, 2018, "Agents Tried to Flip Russian Oligarchs. The Fallout Spread to Trump."

CHAPTER 9: THE PEE TAPE

143 The businessman, David Geovanis: He didn't respond to emails seeking comment.

145 Damian Paletta: He didn't respond to emails seeking comment.

147 Steele secretly dispatched his collector: Steele's collector, Igor Danchenko, told the FBI this in January 2017.

148 Catherine Belton, was a "friend" of his: Belton's piece about Sergei Millian was published in *The Financial Times* on November 1, 2016, "The Shadowy Emigré Touting Trump."

148 Broeksmit soon got a call from Simpson: David Enrich of *The New York Times* described the interactions between Belton, Simpson, and Broeksmit in his 2019 book, *Dark Towers: Deutsche Bank, Donald Trump, and an Epic Trail of Destruction* (HarperCollins).

149 "Don't you have any meeting space": Steele's comments were reported in *The Washington Post* by Tom Hamburger and Rosalind S. Helderman on February 6, 2018, "Hero or Hired Gun? How a British Former Spy Became a Flash Point in the Russia Investigation."

151 It was by Michael Isikoff of *Yahoo! News*: his article about Carter Page appeared on September 23, 2016, "US Intel Officials Probe Ties Between Trump Advisor and Kremlin."

152 passed on business leads: This description of the relationship between Winer and Steele is drawn from the report of the Senate Select Committee on Intelligence.

153 Deripaska's lawyer believes "that the Ukrainians have politicised": Email from Christopher Steele to Bruce Ohr, October 18, 2006.

154 a "Hail Mary pass": In 2019, the Justice Department released memos related to the dossier that Bruce Ohr wrote in 2016.

154 Corn's article appeared a few days after Comey's decision: That article was published on the website of *Mother Jones* on October 31, 2016.

155 "I was completely taken aback": Mike Gaeta made this statement during 2019 congressional testimony during which he was not identified by name. However, that witness's description of his relationship to Steele made it clear it was Gaeta.

155 Steele would insist that he hadn't agreed: Steele made that statement in a press release issued by Orbis Business Intelligence in December 2019, after the publication of the Horowitz report, and in which he criticized several aspects of the report. Among other things, he said he couldn't have agreed as an individual to become an FBI source because of his relationship with the British government.

156 three oligarchs who controlled Alfa Bank: The bank's three founders were Mikhail Fridman, Petr Aven, and German Khan.

157 in anticipation of the *Times* story: This detail was cited by Michael Isikoff and David Corn in their book *Russian Roulette*.

157 *The Intercept*: That article "Here's the Problem with the Story Connecting Russia to Donald Trump's Email Server" appeared on November 1, 2016.

157 when the *Times'* long-awaited "pinging" article: That article by Eric Lichtblau and Steven Lee Myers, "Investigating Donald Trump, FBI Sees No Clear Link to Russia," was posted on October 31, 2016.

159 Carter Page, Paul Manafort, and Michael Flynn: Carter Page was not charged with any wrongdoing. Paul Manafort was convicted of tax fraud charges. Michael Flynn admitted lying to FBI agents about his conversations with a Russian ambassador though he later recanted that admission. George Papadopoulos pled guilty to making false statements to the FBI. In 2020, Donald Trump pardoned Manafort, Flynn and Papadopoulos.

CHAPTER 10: OUTED, EPISODE 1

164 Shell, a Dutch company, and Eni, an Italian firm: As of this writing, prosecutors in Italy were pursuing charges against officials of Shell and Eni. The companies have denied wrongdoing.

166 the Kusto Group: A spokesman for the company denied it had hired K2 Intelligence.

168 "For me, watching this film was a surreal experience": Simon Taylor made those remarks in 2014 while accepting the Skoll Award for Social Entrepreneurship. https://www.youtube.com/watch?v=5n5RNNNCGMs.

168 It was titled *Agent Zigzag*: Ben MacIntyre's book, *Agent Zigzag: A True Story of Nazi Espionage, Love, and Betrayal*, was published in 2007 by Bloomsbury.

169 "It is our strong view that the preferred way forward": Email from Simon Taylor to Rob Moore, August 5, 2014.

170 "For the sake of my family's safety": Email from Rob Moore to Simon Taylor, August 11, 2014.

170 Global Witness had given away the game: Simon Taylor is quoted in my *New York Times* article as saying that it wasn't up to Rob Moore to set the timetable for his unmasking because his undercover actions might jeopardize anti-asbestos activists. "He is not in a position to play God," Taylor said.

171 a reporter affiliated with the International Consortium of Investigative Journalists: That reporter was Jim Morris.

173 the publication *New Matilda*: *New Matilda* published three articles by Michael Gillard about the Rob Moore/K2 Intelligence saga that appeared on March 5, 2017, April 23, 2017, and April 29, 2018.

173 "Brother of BBC boss": The article by Michael Gillard and Jon Ungoed-Thomas appeared in *The Sunday Times* (of London), January 29, 2017.

175 he was starting a new firm: That firm is called C&F Partners.

175 "Charlie Carr has left K2": Email from Mark Hollingsworth to Phillip Van Niekerk, July 10, 2017.

175 *The Irish Times*, published an article in 2016: That article by Peter Murtagh, "Denis O'Brien, the Dossier, and the Spy Who Came into the Dáil," appeared on July 16, 2016.

175 *The Irish Times* reported that Hollingsworth: The newspaper reported that Mark Hollingsworth declined to comment for its article.

176 who now ran a blog: The blog run by Clare Rewcastle-Brown was called Sarawak Report.

176 Jho Low, had sought to set up meetings in 2014: Those efforts are mentioned in June 15, 2017, filing that was part of a forfeiture action brought by the Justice Department against the Viceroy Hotel Group. A spokeswoman for K2 Intelligence didn't respond to an inquiry asking about ties between the firm, Jho Low, or his representatives.

176 Rewcastle Brown wrote on her blog: Email from Mark Hollingsworth to Clare Rewcastle Brown, July 13, 2016.

177 Giannakopoulos and Hollingsworth worked with K2 Intelligence: That article by Bradley Hope and Jenny Strasberg ran on February 26, 2020, "SoftBank's Rajeev Misra Used Campaign of Sabotage to Hobble Internal Rivals." Rajeev Misra denied paying for the operation.

177 stolen from a law firm in Panama: The firm was Mossack Fonseca.

178 "You may recall that we did stories": Email from Mark Hollingsworth to Simon Goodley, February 21, 2017.

178 "Below is a very rough draft": Email from Mark Hollingsworth to Glenn Simpson, April 18, 2013.

179 "Clearly, the story is dependent on PP searches": Email from Mark Hollingsworth to Simon Bowers, March 21, 2017.

179 "My source will not accept anything less than $2,000": Email from Mark Hollingsworth to Richard Hynes, June 16, 2017.

179 "Please email me your hit-list": Email from Mark Hollingsworth to Alexander Yearsley, June 22, 2016.

180 "Since we spoke on the phone tonight": Email from Mark Hollingsworth to Christopher Steele, April 29, 2016.

CHAPTER 11: OUTED, EPISODE 2

182 Ohr's wife: Her name is Nellie Ohr.

182 At the coffee shop, Simpson handed Ohr: Handwritten notes of Bruce Ohr dated December 10, 2016.

183 Kramer started getting a stream of calls from reporters: Testimony of David Kramer in *Gubarev v. BuzzFeed*, dated December 13, 2017.

187 researching a book about the topic: Ken Bensinger's book about the FIFA scandal, *Red Card: How the U.S. Blew the Whistle on the World's Biggest Sports Scandal*, was published in 2018 by Simon & Schuster.

190 He deleted all his files about the dossier: Steele made this disclosure in 2020 during his testimony in the London lawsuit brought by the principals of Alfa Bank. He didn't explain his actions and the timing of them.

192 a "garbage document": Bob Woodward made that remark to Chris Wallace of Fox News on April 21, 2019.

193 Peter Strzok, said in a later interview: He made those comments during an interview in *The Atlantic*, September 4, 2020.

CHAPTER 12: TROJAN WARS

196 known commonly as Pegasus: NSO, the spyware's maker, says it should only be legally used.

196 Citizen Lab's encounter: Raphael Satter of the Associated Press detailed this episode in a January 25, 2019, article, "Undercover Agents Target Toronto-Based Cybersecurity Watchdog Group Which Reported Key Details in Khashoggi Case."

197 he posted them on his personal website: Those aerial photos can be found at https://www.johnscottrailton.com/kite-aerial-photography-kit/.

199 he kept receiving weather alerts: Ronan Farrow wrote about those alerts in *Catch and Kill*.

199 A reporter for a news website, *Motherboard*, exposed the practice: Joseph Cox wrote an article on January 8, 2019, "I Gave a Bounty Hunter $300. Then He Located Our Phone."

201 Raphael Satter, the AP reporter: Satter described his interactions with John Scott-Railton and subsequent encounter with Michel Lambert on January 26, 2019, "Undercover Agents Target Cybersecurity Watchdog."

CHAPTER 13: ROCK STARS

209 the Democracy Integrity Project pulled in more than $7 million: the figure is drawn from group's 2018 990 filing.

210 *The Wall Street Journal* broke a story: Mark Maremont's article about Sergei Millian ran on January 24, 2017, "Key Claims in Trump Dossier Said to Come from Head of Russian-American Business Group."

210 which identified Millian as "Source D": *The Washington Post* piece by Rosalind S. Helderman and Tom Hamburger appeared on March 29, 2017, "Who Is 'Source D'? The Man Said to Be Behind the Trump-Russia Dossier's Most Salacious Claims."

211 "They now have specific concrete": That comment appeared in an April 13, 2017, article in *The Guardian* by Luke Harding and others, "British Spies Were First to Spot Trump Team's Links with Russia."

212 one 2017 show: That episode of *The Rachel Maddow Show* was broadcast on December 29, 2017.

214 Stories about him or Fusion GPS: John Cassidy of *The New Yorker* wrote about Simpson on January 10, 2018, "The Digger Who Commissioned the Trump-Russia Dossier Speaks." Matt Flegenheimer wrote about him for *The New York Times* on January 8, 2018, "Fusion GPS Founder Hauled from the Shadows for the Russia Election Investigation."

216 The lawsuit against BuzzFeed was dismissed based on a judge's finding that the dossier was a government document and thus fair game for publication.

217 *Russian Roulette:* That book by Michael Isikoff and David Corn, which was subtitled *The Inside Story of Putin's War on America and the Election of Donald Trump*, was published in 2018 by Twelve.

217 *Collusion*: Luke Harding's book, subtitled *Secret Meetings, Dirty Money, and How Russia Helped Donald Trump Win*, was published in 2017 by Vintage.

219 "He's James Bond": Rupert Allason's comment appeared in an *NBC News* report, "Christopher Steele, Trump Dossier Author, Is a Real-Life James Bond."

222 a panel discussion at the International Spy Museum: The talk can be found at https://www.c-span.org/video/?c4708289/user-clip-scott-shane-nyt.

222 A former CIA spy on the same panel: The former CIA officer was Jonna Hiestand Mendez.

CHAPTER 14: EPISODE 1: "DOUBLE AGENT"

224 The other report went to an executive at Alaco: Email from Mark
Hollingsworth to Nikos Asimakopoulos, February 23, 2013.

225 In court filings, ENRC said: At the time of this writing, Neil Gerrard's
lawsuit against ENRC and Diligence was continuing. (Black Cube was not
a defendant in the action.)

225 *Intelligence Online*, published an article: The article about the hack of Mark
Hollingsworth's email was published on July 10, 2019, "London's Business
Intelligence Community Braces Itself for Imminent Leak."

226 The email's subject line: The email was dated August 4, 2019.

228 the *Evening Standard*: Mark Hollingsworth's article appeared on September 15, 2017, "SFO Is Stepping Up Its Kazakh Mining Probe."

228 "I have also arranged for this": Email from Mark Hollingsworth to Phillip Van Niekerk, September 15, 2017.

228 "As you know, I am always happy": Email from Mark Hollingsworth to
Rinat Akhmetshin, December 18, 2017.

229 Yuri Shvets: He did not respond to emails seeking comment.

230 made a big reveal: The disclosure that ENRC was paying Mark Hollingsworth and Robert Trevelyan was made by ENRC lawyer, Justin Michael-
son, in a declaration dated December 5, 2019.

230 Robert Trevelyan: He did not respond to emails seeking comment.

230 Dmitry Vozianov: He did not respond to emails seeking comment.

231 "We need very much strong language": Email from Dmitry Vozianov to
Mark Hollingsworth, February 7, 2018.

CHAPTER 15: SHINY OBJECT

233 *The Plot Against the President*: Lee Smith's book, which was subtitled
*The True Story of How Congressman Devin Nunes Uncovered the Big-
gest Political Scandal in U.S. History*, was published in 2019 by Center
Street.

234 *Crime in Progress*: Glenn Simpson and Peter Fritsch's book, which was subtitled *Inside the Steele Dossier and the Fusion GPS Investigation of Donald
Trump*, was published in 2019 by Random House.

235 "Some of the most sensational claims": *The New York Times* article about
the Mueller report by Scott Shane, Adam Goldman, and Matthew Rosen-

berg appeared on April 19, 2019, "Mueller Report Likely to Renew Scrutiny of Steele Dossier."

235 "all but dismissed many key claims": That *Journal* article by Alan Cullison and Dustin Volz appeared on April 19, 2019, "Mueller Report Dismisses Many Steele Dossier Claims."

235 sounded similar themes: Peter Fritsch's comments about the dossier's accuracy were reported on November 22, 2019, by James D. Walsh in *New York* magazine in "Fusion GPS Lights a Candle for the Pee Tape."

236 "The interview was contentious at first": Natasha Bertrand's article appeared on *Politico* on July 9, 2019, "Trump Dossier Author Steele Gets 16-Hour DOJ Grilling."

238 supposed FSB source: This undated memo was sent by Cody Shearer to a journalist.

239 he wrote in an op-ed piece in *The Washington Post*: His op-ed was published on February 8, 2018, "Devin Nunes Is Investigating Me. Here's the Truth."

239 Winer also told: His testimony is contained in the report of the Senate Select Committee on Intelligence.

240 "When you actually get into the details": Michael Isikoff made those comments on John Ziegler's podcast, *Free Speech Broadcasting*, on December 5, 2018.

240 he invited Maddow: Rachel Maddow appeared on Michael Isikoff's podcast, *Skullduggery*, on October 31, 2019.

241 to write a series of columns: Erik Wemple wrote fourteen columns about the media's coverage of the dossier between December 13, 2019, and August 19, 2020.

243 "that Cohen entered the Czech Republic through Germany": That story by Gregg Gordon and Peter Stone appeared on April 13, 2018, "Sources: Mueller Has Evidence Cohen Was in Prague in 2016, Confirming Parts of the Dossier."

243 The same reporters doubled down: Gregg Gordon and Peter Stone published their other Cohen/Prague story on December 27, 2019, "Cell Signal Puts Cohen outside Prague around Time of Purported Russian Meeting."

243 "Trump Campaign Aides Had Repeated Contacts": That article by Michael S. Schmidt, Mark Mazzetti, and Matt Apuzzo appeared on February 14, 2017. (Subsequently, evidence emerged that one Trump aide, Paul

Manafort, worked closely with a suspected Russian operative named Konstantin Kilimnick.)

243 They reported that Paul Manafort: That article by Luke Harding and Dan Collyns appeared in *The Guardian* on November 27, 2018, "Manafort Held Secret Talks wth Assange in Ecuadorian Embassy, Sources Say."

244 they got stiff-armed: The only changes that *The Guardian* made to the article was to include denials from Assange and Manafort that the meeting had taken place. In an email to me, a spokesperson for the paper said it continued to "stand by" the article.

CHAPTER 16: DINNER WITH NATALIA

247 In 2019, the Justice Department had indicted her: Federal prosecutors announced Veselnitskaya's indictment on January 8, 2019. https://www.justice.gov/usao-sdny/pr/russian-attorney-natalya-veselnitskaya-charged-obstruction-justice-connection-civil.

248 Prevezon agreed to pay a $5.9 million fine: The announcement of the case's settlement was made on May 12, 2017. https://www.justice.gov/usao-sdny/pr/acting-manhattan-us-attorney-announces-59-million-settlement-civil-money-laundering-and.

250 a bipartisan Senate report: This was the report of the Senate Select Committee on Intelligence.

251 a decision Akhmetshin had appealed: In late 2020, a federal court granted his appeal and sent the case back to a trial court for reconsideration.

252 Claims filed by ENRC against Mark Hollingsworth and others: At the time of this writing, ENRC's lawsuits against Mark Hollingsworth, the Serious Fraud Office and others were continuing. Defendants in all the actions have denied any wrongdoing.

253 In a lengthy article in 2018: Dexter Filkins's article in *The New Yorker* ran on October 8, 2018, "Was There a Connection Between a Russian Bank and the Trump Campaign?"

254 In his testimony, Steele appeared to display a consistent attitude: Steele testified in the Alfa Bank–related case on March 17, 2020.

255 the judge overseeing the case issued his ruling: That ruling was issued on August 7, 2020. https://www.casemine.com/judgement/uk/5f06a0562c94e070322b31e6.

256 "We were no longer in the contract with Fusion": Steele testified in the Gubarev case on July 23, 2020.

CHAPTER 17: THE COLLECTOR

262 when those interview notes hit the internet: The notes of Igor Danchenko's interviews were released by the Justice Department on July 17, 2020.

264 He said during his FBI interviews: Igor Danchenko was interviewed over the course of three days in late January 2017.

266 Luke Harding: He did not respond to emails seeking comment about this discrepancy.

269 Igor Danchenko: His lawyer Mark Schamel didn't respond to written questions.

270 He told Luke Harding: Igor Danchenko's interview appeared in *The Guardian* on October 21, 2020, "Trump's False Russian Spy Claims Put Me in Danger."

270 published on the same day: Danchenko's interview in the *Times* on October 21, 2020, "Analyst Who Reported the Infamous Trump Tape Rumor Wants to Clear His Name."

271 issued a long-awaited report: The fifth volume of the Senate Select Intelligence Committee's report was issued on August 18, 2020.

272 a documentary: The name of Alex Gibney's two-part film was *Agents of Chaos*.

276 arrested an Israeli-based private investigator: At the time of this writing, the charges against that investigator were still pending.

INDEX

ABOUT THE AUTHOR

BARRY MEIER is a former *New York Times* reporter and a member of the *Times'* team that won the 2017 Pulitzer Prize for International Reporting. He is also a two-time winner of the prestigious George Polk Award for Investigative Reporting and other professional honors. Prior to joining the *Times* in 1989, he worked for *The Wall Street Journal* and *New York Newsday*. He is also the author of *Pain Killer* and *Missing Man*. Meier lives in New York City.